BEYOND CARTESIAN DUALISM

Science & Technology Education Library

VOLUME 29

SCOPE

The book series *Science & Technology Education Library* provides a publication forum for scholarship in science and technology education. It aims to publish innovative books which are at the forefront of the field. Monographs as well as collections of papers will be published.

The titles published in this series are listed at the end of this volume.

Beyond Cartesian Dualism

Encountering Affect in the Teaching and Learning of Science

Edited by

STEVE ALSOP
York University, Toronto, Canada

Springer

A C.I.P. Catalogue record for this book is available from the Library of Congress.

ISBN-10 1-4020-3807-0 (HB)
ISBN-13 978-1-4020-3807-5 (HB)
ISBN-10 1-4020-3808-9 (e-book)
ISBN-13 978-1-4020-3808-2 (e-book)

Published by Springer,
P.O. Box 17, 3300 AA Dordrecht, The Netherlands.

www.springeronline.com

Printed on acid-free paper

Printed in the Netherlands.

TABLE OF CONTENTS

ACKNOWLEDGEMENTS

It is a pleasure to acknowledge and congratulate all those involved in the evolution of this book. First and foremost, I would like to thank the authors. The text has been gestating for some time and they have been extremely patient and supportive as the collection took shape. I have learned much from reading their work and our ongoing email exchanges. English language is not the first language of some of the authors and I am particularly grateful for their tolerance and trust in my editing. I would especially like to thank Bill Cobern for his vision, expertise, and prompt encouraging emails. Bill played a key role in keeping the editing process going especially at times when my faith in the book was wavering.

I am indebted to Sheliza Ibrahim for her thorough, insightful editorial expertise. Marion Wagenaan, Bernadette Deelen, Gerrit Ooman and Arvind Sohal at Kluwer–Spinger have been supportive and understanding throughout. They kindly helped me with the typesetting and indexing stages. Three reviewers kindly read a first draft and their perceptive and challenging comments helped to restructure the text and make it better in a number of ways.

The idea for the text emerged in a series of conversations with Mike Watts, and I would like to express my gratitude to him for years of collaboration, stimulating conversations, provocative ideas, and invaluable mentorship, advice, and guidance.

It should not go without saying that this project would not have been possible without Vanessa's, Dylan's, and Olivia's encouragement.

Like many academics, my thinking in the area of motivation has been greatly influenced by the seminal work of Paul Pintrich. I was both delighted and humbled that Paul displayed such an interest in the evolution of this collection. Indeed, Paul was one of the first academics that I contacted to see if he would be interested in crafting a chapter. He replied within minutes.

Paul's chapter is published here posthumously. I thank colleagues at the University of Michigan, Christina Rhee Bonney, Toni Kempler, Akane Zusho, and Brian P. Coppola, for their willingness to continue with this project in extremely difficult circumstances. Paul's ideas are to be found not only in this excellent chapter but also weave their way through the entire collection. My hope is that this serves as a testimony to a truly special, dedicated, and highly influential scholar.

CONTRIBUTORS' DETAILS

In order of appearance:

Steve Alsop is Associate Dean (research, international, and teacher professional development) in the Faculty of Education at York University, Canada and Senior Honorary Research Fellow at Roehampton University, United Kingdom. Steve has taught in primary schools and secondary schools in the United Kingdom and previously coordinated CLARISE, the Centre for Learning and Research in Science Education (CLARISE). He has a wide range of publications in science education, including two previously edited books [Alsop, S., & Hicks, K. (Eds.). (2001). *Teaching science*. London: Kogan Page; Alsop, S., Bencze, L., & Pedretti, E. (Eds.). (2005). *Analysing exemplary science teaching: Theoretical lenses and a spectrum of possibilities for practice*. Milton Keynes, United Kingdom: Open University Press].

Michael Reiss is Professor of Science Education at the Institute of Education, University of London and Head of its School of Mathematics, Science, and Technology. He is Chief Executive of Science Learning Centre London, Honorary Visiting Professor at the University of York, Docent in Science Education of the University of Helsinki, Director of the Salters-Nuffield Advanced Biology Project and editor of the journal *Sex Education*. For further information see www.reiss.tc.

Alice Pitt is Associate Professor of Education and the Associate Dean of Pre-Service Education in the Faculty of Education at York University, Toronto, Canada. Recent publications include *The Play of the Personal: Psychoanalytic Narratives of feminist Education* (2003, Peter Lang); On losing and refinding the mother: Reading women's autobiographies. *Changing English: Studies in Reading and Culture, 11*(2), 267–277 (2004); and with Deborah Britzman, Speculations on qualities of difficult knowledge in teaching and learning: An experiment in psychoanalytic research. *International Journal of Qualitative Studies in Education, 16*(6), 755–776 (2003).

Martina Nieswandt is an Assistant Professor of Science Education at the Ontario Institute for Studies in Education of the University of Toronto (OISE/UT). She graduated from the University of Kiel, Germany, and worked as a researcher at the Leibniz Institute for Science Education (IPN) in Kiel, Germany, prior to working in North America. Her current research projects investigate the relationship between the affective, motivational, and cognitive dimensions on learning high-school science. Furthermore, she is interested in challenging preservice and inservice science teachers' epistemological beliefs about science and science teaching through her involvement in preservice science teaching and research projects with classroom teachers.

Camilla Schreiner has her master in geophysics and is currently doing her PhD on the ROSE project at the University of Oslo. Her research interests are affective qualities in science and environmental education, cross-cultural studies and sociological perspectives on modernity and youth culture.

Svein Sjøberg is Professor in Science Education at Oslo University. His research interests include: Social, cultural, and ethical aspects of science education, science education and development, gender and science education in developing countries, critical approaches to issues of scientific literacy and public understanding of science. He is involved in several international organizations and initiatives, and member of the Expert Advisory Committee on the European Science & Society Action plan for ethics, gender equity, and public dialogue under Frame Programme 6. Information and articles on http://folk.uio.no/sveinsj.

Sherry A. Southerland is an Associate Professor of Science Education at Florida State University. She received her PhD in Curriculum & Instruction from Louisiana State University with her research focusing on students' conceptual ecologies for biological evolution. Her current research interests are the role of conceptual ecologies in conceptual change, the influence of culture on science learning understanding the intersection of nature of science knowledge, and the learning of controversial topics. She is currently a coordinator for strand 3, science teaching, for *NARST.*

Gale M. Sinatra is a Professor of Educational Psychology at the University of Nevada, Las Vegas. She received her PhD in Psychology from the University of Massachusetts in 1989 where she specialized in reading and measurement from a cognitive perspective. Her current research interests include conceptual change learning, epistemological beliefs and learning dispositions, and intentional learning. Her recent book, *Intentional Conceptual Change*, was co-edited with Paul Pintrich, and published by Lawrence Erlbaum Associates. She is currently serving as Secretary of Division C, Learning and Instruction, of the American Educational Research Association and Editor of the American Psychological Association Division 15 journal, *Education Psychologist*.

Christina Rhee Bonnie received her B.A. in psychology from Yale University and her M.S. in developmental psychology from the University of Michigan. She is currently a doctoral candidate in the Combined Program in Education and Psychology at the University of Michigan. Her research interests include investigating students' and athletes' achievement goals and learning strategies, and the influence that the perceived achievement context has on their subsequent motivation and performance.

Toni M. Kempler is currently a doctoral candidate in the Combined Program in Education and Psychology at the University of Michigan, Ann Arbor. She received her B.A. in elementary education and psychology from Goucher College and her M.S. in developmental psychology from the University of Michigan. Her research examines the influence that the classroom context has in promoting motivation, cognitive engagement, and achievement. She has studied the relation of motivation to students' strategy use, emotional well-being, and learning within the context of collaborative groups in mathematics. Her current work is grounded in standards-based reform efforts that aim to enhance student engagement and achievement in the areas of science and mathematics.

Akane Zusho is currently an Assistant Professor of Educational Psychology in the Graduate School of Education at Fordham University. She received her B.A. in psychology, her M.A. in developmental psychology, and her Ph.D. in Education and Psychology, all from the University of Michigan. Her program of research focuses on examining the development of academic and motivational processes of urban youth from culturally diverse backgrounds.

Brian P. Coppola is Arthur F. Thurnau Professor of Chemistry at The University of Michigan, and a Faculty Associate at the University of Michigan Center for Research on Learning and Teaching. He currently serves as the department's Associate Chair. Dr. Coppola received his B.S. degree in 1978 from the University of New Hampshire and his Ph.D. in Organic Chemistry from the University of Wisconsin–Madison in 1984, having joined the faculty at the University of Wisconsin–Whitewater in 1982. Moving to Ann Arbor in 1986, he joined an active group of faculty in the design and implementation of a revised undergraduate chemistry curriculum. His 1996–1997 tenure review established a new policy within the College of Literature, Science, and Arts at the University of Michigan, recognizing discipline-centered teaching and learning as an area that can be represented within the LSA departments. He was promoted to Full Professor of Chemistry in 2001–2002. His recent publications range from mechanistic organic chemistry research in 1,3-dipolar cycloaddition reactions to educational philosophy, practice, and assessment.

Paul R. Pintrich was a Professor of Education and Psychology and Chair of the Combined Program in Education and Psychology at the University of Michigan, Ann Arbor. He had a PhD in Education and Psychology from Michigan, an M.A. in Developmental Psychology from Michigan, and a B.A. in Psychology from Clark University. He had published over 100 articles, book chapters, and books on motivation, self-regulated learning, and adolescent development. Dr. Pintrich had served as the President of Division 15—Educational Psychology of the American Psychological Association as well as President of Division 5—Educational and Instructional Psychology of the International Association of Applied Psychology. He also served as the editor of the journal, *Educational Psychologist*, from 1995 to 2001.

Jerry Wellington is currently Head of Research degrees in the School of Education, University of Sheffield, United Kingdom. He taught science in London schools before moving to Sheffield. He is author of over 70 publications, including the following books: Wellington, J. J. (2000). *Educational research: Contemporary issues and practical approaches.* London: Continuum; Wellington, J, & Osborne, J (2001). *Language and literacy in science education.* Buckingham: Open University Press; Wellington, J. (2003). *Getting published.* London: Routledge.

Lynn D. Dierking is internationally recognized for her research on the behaviour and learning of children, families, and adults in free-choice learning settings such as museums and has published and spoken extensively in these areas. She possesses a PhD in Education from the University of Florida, Gainesville, and her research priorities include the long-term impact of free-choice learning experiences on individuals and families and the development and evaluation of community-based programs. Over the last 20 years Dr. Dierking has worked in a variety of settings, including: the Smithsonian Office of Educational Research, University of Maryland's College of Education and as director of a national curriculum project, Science in American Life, at the Smithsonian's National Museum of American History. Her publications include three seminal books co-authored with Dr. John H. Falk, *The museum experience* (Whalesback Books, 1992), *Learning from museums: Visitor experiences and the making of meaning* (AltaMira, 2000), and *Lessons without limit: How free-choice learning is transforming education* (AltaMira, 2002. She also co-edited a volume, *Public institutions for personal learning: Establishing a research agenda,* with Dr. John Falk in 1995.

Michalinos Zembylas is Associate Professor of Education at Intercollege, Cyprus, 46 Makedonitissas Avenue, Nicosia 1700, and adjunct professor of teacher education at Michigan State University. His research interests are in the areas of science and technology education, curriculum theory, and philosophy of education. His recent articles have appeared in *Educational Theory*, *Journal of Research in Science Teaching*, *Science Education* and *Teaching and Teacher Education*.

Federic Perrier is a physicist and physics educator. His previous positions include a postdoctoral research associate at Stanford University (United States) and a research scientist at *Commissariat énergie atomique*, France. Currently he is an Associate Professor at *Institute de Physique du Globe de Paris* (a position he has held since 2003). For over a decade, Federic has been involved in the rehabilitation of underprivileged children in Nepal and Rwanda. He implemented the WASP (Workshops of Active Science for Peace) project in Rwanda in 2000 and is a part-time science teacher in Nepal.

Mike Watts is Principal of Froebel College and Professor of Science Education at Roehampton University. He teaches at all levels within the university and directs several research projects in learning and teaching. He is widely published in the field of science and technology education, in teaching and learning in higher education and is the author of numerous works on educational research and research methods. He is a National Teaching Fellow of the Higher Education Academy, and a Fellow of the Institute of Physics.

Bonnie Shapiro is a Professor in the Faculty of Education at the University of Calgary holding joint appointments in the Divisions of Teacher Preparation and the Graduate Division of Educational Research. She is a former classroom teacher with interests in studying the culture and demands of living and working in learning environments. She gives high status to and tries to understand and describe the intellectual, social, and emotional demands of learning. Some of her original research interests have focused on children's construction of meaning in science (see Shapiro (1994) *What children bring to light: A constructivist perspective on children's learning in science*), which has become a longitudinal study of children's learning and valuing of science in their lives. Engagement in this work has naturally led her to an interest in how preservice and in-service teachers interpret the challenges and meaning of their work with children.

Brian Matthews taught science in secondary schools in London for 19 years and is now at Goldsmiths College where he trains science teachers on the PGCE. He is interested in finding ways of increasing pupils' interest in science and of enabling boys and girls to relate well together. Brian's research interests are all connected to equity issues and include the development of emotional literacy and using pupil's drawings of scientists to encourage reflection about science. His recent book Engaging Education. Developing Emotional Literacy, Equity and Co-education, Open University Press, is the first book integrate social justice with emotional literacy.

INTRODUCTION: SCIENCE EDUCATION AND AFFECT

CHAPTER 1

STEVE ALSOP

BRIDGING THE CARTESIAN DIVIDE: SCIENCE EDUCATION AND AFFECT

INTRODUCTION

There is surprisingly little known about the emotional aspects of science education. A senior scholar, over a decade ago, called for a fundamental shift in our field, suggesting that the "affective area will prove to be crucial, in research and curriculum planning in the future" (Head, 1989, p. 162). However, despite periodic forays into monitoring students' attitudes-towards-science, the "effect of affect" in the teaching and learning of science continues to be largely ignored.

Existing research, while sparse, mostly presents a rather gloomy picture. After decades of research and curriculum reform, sources indicate that while some elements of science education engender fascination and awe, too many middle and secondary school students in high-income countries find school science overly mundane and lacking relevance (Sjoberg, 2002). Indeed, recent work suggests that attitudes and post-compulsory involvement in science education (particularly by women and ethnic minorities) are still on the decline (Osborne et al., 2003). But while research has been successful in defining the problem, it has yet to say very much about the solution, and this clearly needs to change.

In broad terms, this is the challenge taken up here: to collect together contemporary theorising about the relationship between affect and cognition in science education and explore the implications that this has for further research and pedagogy. It is the centrality of affect and its lack of consideration that has driven my desire to compile this edited collection. After all, there is overwhelming evidence from a diversity of fields including Psychology (Sutton & Wheatley, 2003), Philosophy (Goldie, 2002), Cultural Theory (Shweder & LeVine, 1986), Feminist studies (Boler, 1999) and Neuroscience (Demasio, 2000, 2004) and Science Education itself (see Alsop & Watts, 2003a; Falk & Dierking, 2000; Zembylas, 2002, 2004), that affect and cognition cannot be meaningfully understood as disparate entities.

Emotions, after all, have considerable influence over what happens in the classroom. Some (such as, joy, love, happiness, and hope) act to enhance education, optimise student enjoyment and achievement. Here to use the language of Csikszentmihalyi (1988, p. 127) teachers and learners become swept up in a world of consciousness, a "*flow* experience," describing themselves as being "carried away by a current"; existentially lost in thought. Now, education is more than the memorisation of a curriculum subject, the anesthetised acquisition of a remote object. It is the beauty and delight of becoming absorbed, seeing the world in different ways with different possibilities. It is

Steve Alsop (ed.) Beyond Cartesian Dualism, 3–16.

about challenge, surprise, desire, joy, expectation, and mystery: the thrill of discovering oneself in relation to new ideas and contexts. When we teach, we invite students into the beauty of our worlds, "to embrace what we have found so alluring" (Liston, 2004, p. 101); to understand the promise, mystery, and intrigue associated with a subject that has occupied our hearts and minds for so long.

But unfortunately, this promise of education can be lost. Unchecked daily practices can inadvertently overwhelm thinking and close down concentration, such that learners' efforts are swamped and rendered wholly ineffective. Claxton (1989, p. 155) writes of turbulent flow; "cognition doesn't matter if you're scared, depressed or bored." The risk of exposure and the potential for humiliation serve to deaden curiosity and insight. Discussion of patterns and complexity acts to do little more than instil pangs of inadequacy as fledgling scientists become crushed by the conjectures of giants, the theories of Newton, Darwin, Einstein, Schrödinger, Priestley, and others (Alsop, 2001b). The list is indeed daunting. For some learners, the fragile formation of an identity within science becomes lost as they withdraw from the ring and retreat to the comfort and security of the familiar, rejecting the uncertainty and risk associated with "border-crossing" into the subculture of science (see Giroux, 1992 and Aikenhead, 1996). What emerges now is a sense of distance, a broken attachment with a subject that becomes all too readily dismissed as "too difficult," "not for me," and "too boring." The concept of "science-for-all" becomes little more than empty rhetoric, as students (more often than not, women and ethnic minorities) opt out of a subject that they see as incommensurate with their interests, aspirations, and future needs.

Education works best when it combines hearts and minds. As Dewey (1931, p. 189) writes, "there is no education when ideas and knowledge are not translated into emotion, interest and volition." Teachers, I believe, have known this for a long time. The following chapters assert that it is time for scholarship in our field to allow, perhaps even invite, affect into the academy as a way of better understanding (and enhancing) science education.

This book I believe to be the first of its kind; a volume gathering-up contemporary theorising about the role of feeling and emotion in the teaching and learning of science. Quite why it has taken so long for a project with this focus to emerge is possibly revealing. Theorists might relish the opportunity of exploring sustained inattention, how individual practices and sociocultural forces have aligned to suppress affect, and perpetuate the myth that studying science is the most traditional, rationalistic, and challenging of all academic endeavours.

BUILDING AN EDITED COLLECTION

In building an academic collection, the absence of a monolithic theoretical framework carries the advantages of freedom but feels naked without tradition. As affect and science education is theoretically youthful, I enjoyed considerable flexibility in my role of editor. From the outset it was never my intention that contributing pieces would capture a central claim; scratch out a unified theory which might serve to funnel further agendas. Neither was it my plan to represent the emphasis of existing work. The substantive work in science education and affect has been on exploring "attitudes-towards-science." A selected review of this work is to be found in the following section in Chapter 4. But rather than an extended discussion of attitude, the following chapters proceed on a more exhaustive front. They seek to bring to the fore a host of different theoretical frameworks and methodological perspectives. Emotion, after all, has been defined and studied in a multiplicity of different ways. Cornelius (1996), for instance, outlines four theoretical traditions (Darwinism, Jamesian, Cognitive, and Social Constructivist). Damasio (2004) and Freud offer others (neurological and psychoanalytical perspectives, respectively). The literature is indeed vast. Here, the intent is not to represent the mass of this scholarship, but instead to provide the reader with a modest selection of theorising that has proved efficacious in the field.

About a year ago, a Special Issue of the *International Journal of Science Education* focused on affect. A mentor and friend, Professor Mike Watts and I (Alsop & Watts, 2003b) acted as guest editors. The overwhelming response to the call for papers provided the impetus to bring together a more extensive collection. To invite scholars positioned in the vanguard to share some of their work and to signpost future empirical and pedagogical directions. All the articles herein are at the forefront of current theorising. They are original pieces insofar as they have not been published elsewhere; although some of the empirical work, as one might imagine, has (and will) feature in future articles and papers.

THE STRUCTURE OF THE BOOK

In the chapters that follow, you are introduced to existing and emerging work exploring the role of affect in science education. These discussions are presented in an introductory section, followed by three main sections (comprising a total of 14 chapters). At the start of each section, I provide a brief overview of the proceeding chapters.

The first section, "Students' Attitudes, Hopes and Dispositions," houses three chapters. In these, scholars press extant literature and new empirical data in search of answers to a wide range of questions, including

- What is meant by the affective constructs; attitudes, hopes, intentions, and dispositions? How might they be measured? What are science students' attitudes, hopes, and dispositions? How do these constructs associate with gender, instructional strategies, curriculum areas, knowledge, acceptance, and student achievement?

The discussions have an individual emphasis. Affect, in this sense, is seen as a psychological construction manifesting itself in individual's expressions, often captured by qualitative research methods, such as surveys and Likert scales. The chapters are arranged progressively and shift from a more general typical consideration of attitudes-towards-science to specific consideration of how the notion of "controversy" in a scientific subject might influence learning, and conclude with environmental challenge and hopes for the future.

In the second section, "Teaching, Learning, and Affect," discussions delve into the learning environment and explore contextual features and teachers' emotions. A series of quite different questions now emerge, including

- How does affect mediate and moderate instruction? How might instructional strategies be better understood by exploring affect? How might free choice learning situations offer a unique context in which to chart affect?

Here, reviews of the literature and data are used to better understand teaching and learning by recognising the influence of emotion within the learner/environment dialectic. In most cases, the overarching goal is to increase cognition, to understand how particular instructional practices might serve to increase motivation, and as a consequence enhance learning and achievement. In this section, the first three chapters explore motivation and swing from an exploration of pedagogy and extrinsic motivation in classrooms to "free choice" learning and intrinsic motivation in museums.

The concluding chapter, Chapter 10, changes focus and theoretical emphasis. From the perspective of poststructural and feminist frameworks it evaluates contemporary scholarship exploring teachers' emotions and proposes some future direction for work in this growing area.

In the third section, "Pedagogical Interventions," the authors describe a series of situated case studies of practices that have sought to purposefully incorporate affect. The collected chapters explore affect in a wide range of educational settings using a diversity of empirical approaches. They explore an array of questions, including

- How might practice more effectively engage learners? How might science education act to restore psychosocial resilience? How might science education engage affect as a way of challenging inequality? How are the pre-service teachers' emotions related to their classroom identity?

In more general terms, these sections look to model the relationship between cognition and affect in science education in different ways. The evolving arguments, although far from theoretically homogenous, we hope are persuasive, cumulative, and generative. They come together here to challenge a lingering 17th century tradition; offering a vision of science education—as the title suggests—*Beyond Cartesian Dualism*.

THE LEGACY OF DUALISM

From Ancient Philosophy and the rise of Greek Civilization (Socrates, Plato, and Aristotle), to the Renaissance (Hume, Kant) and the 20th century, few scholars would overlook the importance of Rene Descartes. Bertrand Russell (1947), for instance, labels Descartes as the founder of modern science. His mechanical philosophy certainly brought into question the relationship of mind and matter and offered an uncomplicated and unencumbered way of severing the body from the mind. Descartes' brilliance was his ability to weave together two worlds that had hitherto remained separate, the physical–inorganic and the living–organic. One consequence, however, of the Cartesian revolution was the loss of emotions, which became seen as part of the body—separated and relegated from the rarefied rationalism of the mind. Emotional responses such as sorrow and joy are discussed at length in the *Mediations* and conveniently (if not persuasively) separated from more rationalistic considerations of hunger and thirst. Cartesian dualism offered a convenient means of elevating the pureness of reason from the messiness and irrationalism of emotion. Of course, such views are no longer mainstream in science and philosophy, but my suggestion is that their remnants are to be found in contemporary science education research and policy reform.

Take for instance the science curriculum. In schools and universities in its prepositional form or as curriculum-in-practice, it is largely portrayed as affect free. School texts, more often than not, present anodyne images of science and scientists—images that often embody technicality, rationalism, and emotional aloofness (Alsop & Watts, 2003b). Auguste Comte's doctrine of positivism, although largely dismissed in the history and philosophy of science, emerges unscathed in policy that reinforces the split between facts and values (see Hodson, 1998). In a compelling Cartesian tradition, many curricula focus entirely on scientific attitudes (the rationality of inquiry, hypothesising, experimenting, and concluding) and completely ignore attitudes-towards-science (the emotions associated with studying school science and those associated with science itself; see Simpson et al., 1995). Such an approach arguably serves to separate the proficiencies of science and technology, *learning to do science* (Hodson, 1998), from the emotions of mystery, creativity, and intrigue. It acts, albeit perhaps inadvertently, to underscore a dichotomy between Socratic assumptions (rational scientific thought in pursuit of "objective" knowledge) and emotions (inherently irrational considerations).

The persistence of the "third–person" laboratory write-up, perhaps more than anything else, serves as a beacon to dualism. There is a significant difference between talking and writing about something in the "third person" and writing about something impersonally. It is possible, of course, to write in the third person while expressing a personal point of view. However, this distinction was certainly lost in my school and university education, in which the third-person language of the laboratory was introduced as the authentic scientific means of recording depersonalised observations, reinforcing objectivity and suppressing the mention of feelings.

In a more general sense, there is a suspicion in Western culture that there is something wrong with emotions. Emotions are irrational processes that lead to weakness and vulnerability and need to be closely monitored and ultimately controlled. Descartes (1637/1911) portrayed emotions as passions of the soul, emphasising their passive, oppressive nature rather than their potential for liberation. Even though we no longer think about emotions in his narrow way, this view is embodied in our everyday language and culture. We talk about being "gripped by fear," "overcome with embarrassment," "seized by anger," "paralysed by anxiety," "overtaken with joy" (see Cornelius, 1996, p. 156). Such metaphors represent emotions as things that are obstructive and beyond our immediate control—manifestations of self-oppression rather than "instruments of freedom" (see de Sousa, 1980). Given this, it is perhaps not surprising that researchers who pride themselves on rationality and logic have been slow to embrace the affective domain.

As previously mentioned, there is certainly a paucity of work in science education that has explored affect. The tradition is cognition. Piaget's epistemic entity and the children's science movement (Gilbert et al., 1982), for instance, challenged the fallacy of discovery learning and the simplicity of treating children as empty knowledge vessels, *tabla rasas*. Researchers fervently studied the ways in which constructs are experienced and conceived by learners, but for the most part, this research presented learning as conceptual change—an epistemological, rationalistic process. The Conceptual Change Model (CCM) of Strike and colleagues (1982) is a case in point. The model was extremely influential and spawned a wealth of research and classroom innovation (Desmastes et al., 1996; Smith, 1991; Songer & Mintzes, 1994). A decade later, it was modified in response to critique offered by Pintrich and colleagues (1993), but even in its revised format (Strike & Posner, 1992) emotions were still given a subsidiary role, relegated to the broad and nebulous notion of a "conceptual ecology."

The commitment was to rational processes in which *intelligibility, plausibility,* and *fruitfulness* dominated as factors determining conceptual change. Even Matthews' (1995) vociferous and widely publicised critique of the children's science movement focused on its "empirical naivety" rather than its cognitive emphasis.

More recently, scholarship has gravitated towards socioconstructivist models of learning accentuating the dynamic dialectic between the learner and the social and cultural context (Vygotsky, 1986). But again, while this perspective might have served to broaden the agenda by recognising the close interaction between the cognitive, social, and emotional (Zembylas, 2004), the vast majority of theorising has favoured the exploration of cognitive and meta-cognitive processes. Vygotsky (1978, 1986), throughout his writing, actually rejected Descartes' notion of dualism and yet, it is the distilled cognitive reservoirs of his scholarship that shape our contemporary theorising. Empirical work, for example, has charted the cognitive boundaries of the Zone of Proximal Development (ZPD) and pedagogical practices that are linked directly with cognitive outcomes. Only very recently has the notion of *emotional* scaffolding emerged. Rosiek and colleagues at the University of Stanford offer the following definition:

> Teachers' pedagogical use of analogies, metaphors, and narratives to influence students' emotional response to specific aspects of the subject matter in a way that promotes student learning. (Rosiek, 2003, p. 402)

In an educational climate of measurable cognitive outcomes, high stakes testing and external accountability measurements, all too often—it seems—educational research and practice looks to denigrate emotion. Early work on teachers' lives is another case in point. These studies focused almost exclusively on weakness and fragility and the resulting narratives fixated on teacher fatigue and burnout, casting affect in a negative light, as an obstacle—an approach that served to undermine the existence of a "caring" practitioner (Ashforth & Humphrey, 1995). Dualism was, once again, re-enacted in a dialogue that framed affect as the "other"—a hindrance, a countenance to reason and informed action.

Thankfully, this view is now changing and this edited collection points to a broader and more thoughtful discussion of the emotional life of teaching (see Zembylas and Shapiro in Chapters 10 and 13 respectively). Hargreaves (1998, p. 835), for instance, has championed the return of affect by writing about the "heart of teaching being charged with positive emotions." Good teachers, he continues, are "emotional, passionate beings who connect with their students and fill their work and their classes with pleasure, creativity, challenge and joy." The emotional practice of teaching is analysed within complex social interactions of schooling, the relationships amongst teachers, students, parents, and administrators. Although, this work seems somewhat at odds with a recent study conducted by Tobin and McRobbie (2001) suggesting that science teachers commonly make sense of their practice in terms of four mechanistic myths: (a) the transmission of knowledge, (b) being efficient, (c) the rigor of the curriculum, and (d) student accountability.

For some time, feminist scholars, in particular, have challenged the legacy of dualism embedded within Western patriarchal thought, questioning the political motivations separating the rational from the irrational. Boler (1999, p. ix), for instance, writes about reading "between the lines" to find mention of emotion and how she was drawn to the conclusion that dualism was far from a coincidence and oversight. She critically laments

> The boundary—the division between "truth" and reason on the one side, and "subjective bias" and emotion on the other—was not equal: emotion had been positioned on the "negative" side of the binary division. And emotion was not alone on the "bad" side of the fence—women were there too. (p. xv)

Studies of attitudes (see Chapter 4) continue to emphasise that girls' attitudes towards science are significantly less positive than boys. Although it is almost impossible to overestimate the importance and influence of existing scholarship in science education and gender, it is perhaps revealing to note that this work has been largely drawn to considerations of epistemology and sociocognition rather than emotion *per se* (see Baker, 2002; Harding, 1991; Kahle, 2004). Perhaps affect might offer a different vantage point and further insight into this critical debate.

So, to borrow the words of the neuroscientist Damasio (2004), this collection seeks, in very broad terms, to illuminate "Descartes' Error." The reviews and empirical studies gathered together act to underscore the complexity of science education, a complexity that necessitates recognition of the mutually constitutive nature of cognition and affect. The authors challenge the problem of the verisimilitude embedded within the notion of the individual mind that acquires a mastery of science in an emotional, social, and cultural vacuum, and render visible the dilemma of considering affect as an obstacle, a barrier to reason and enlightenment. The affective dimensions of cognition are heeded not as a way of helping disillusioned, disenfranchised learners overcome their irrational fears in a quest for conceptual enlightenment. At all levels, cognition and affect are seen as fused, inseparable. Pedagogy should not look to tame and sterilise feelings, effectively dulling creativity and sanitising imagination. Instead, affect should be seen as axiomatic, at times making science education difficult, but above all else, actually making science education *possible*.

BEYOND THE CARTESIAN DIVIDE: FUTURE DIRECTIONS FOR RESEARCH

In this concluding section, I offer some reflections on future research. My thoughts are, in part, drawn from the following chapters (many of which contain specific recommendations for future scholarship) combined with recent publications, which I suggest might prove influential in the future.

In a very broad sense the discussions herein urge a re-examination of existing deficits as well as encouraging future possibilities. They invite a critical exploration of exclusionary habits as well as an investment in the challenge of understanding how emotion might play a key role in the field. In so doing, they raise some general questions for future theorising, including:

> What are the existing assumptions of science education? How might we avoid learners retreating from our subject? How might future studies actively explore and "problematise" the under-representation of affect? In what ways might we promote emotional investment in science education? What happens when science educators plan pedagogical approaches that prioritise emotion?

The collected work offers future scholarship a number of different affective referents and theoretical frameworks to explore the relationship between cognition and affect. These include learners' attitudes, beliefs and hopes, as well as models of conceptual change (see Chapters 4, 5, and 6); psychoanalytical models of teaching and learning (see Chapters 2, 3, and 11); motivational beliefs (See Chapters 7, 8, and 9); socioconstructivist models and poststructuralist conceptions of teachers' emotions (Chapter 10); and a series of empirical case studies of affect in context (see Chapters 12, 13, and 14). These theoretical commitments raise a variety of questions for future science education research. Some of these are listed below within the areas of Learning Science, Teaching Science and Science Teachers' Emotions.

Learning Science

From a constructivist perspective, future scholarship might look to extend established models of learning. For example, revisions of the aforementioned CCM (Strike & Posner, 1992) now exist that have been modified to embrace affect. My early work, offered an Extended Conceptual Change Model (ECCM) comprising cognitive, affective, and conative elements (see Alsop, 1999). Perhaps more progressively, Dole and Sinatra (1998) have proposed the Cognitive Reconstruction of Knowledge Model (CRKM) and Gregoire (2003) has put forward the Cognitive-Affective Model of Conceptual Change (CAMCC). The CAMCC, as Southerland and Sinatra explore in Chapter 5, has the advantage that it models the way in which affect might direct (rather than influence) cognition. Future work might build on and revise these models and seek to investigate a variety of questions, including

> What types of emotions do learners experience? What emotions does conceptual change engender? What emotions are associated with cognitive conflict and conceptual dissatisfaction? How do these emotions influence and/or direct conceptual change? Might different science content engender different emotions? How might learners' beliefs and attitudes mediate and moderate conceptual change? How might controversy make learning easy or difficult?

As the following chapters demonstrate, different scientific topics are likely to arouse different emotions and as a consequence they might be best learned and taught in different ways. A theory of content, as White (1994) suggests, is probably long overdue. Few would disagree that some science content is provocative and when encountered is more likely to arouse intense emotions. Although we have a track record of monitoring learners' attitudes towards science, we actually know very little about the emotions associated with particular aspects of science. A finer grained analysis might investigate, for instance

> What emotions do learners associate with science study? What aspects of science do learners find controversial? What might be considered emotionally difficult scientific knowledge? What images of science do learners find appealing? What images do learners find distasteful?

My own work (Alsop 1999, 2001a) has sought to model the role of controversy and fear, within the context of the teaching and learning of radioactivity. More recently, Zembylas (2004) has investigated children's emotional practices while engaged in long-term science investigations. These studies might offer future directions.

In the past, theory and practice has largely focused on how learners acquire knowledge, rather than exploring learners' and teachers' relationship with knowledge. Here, perhaps the distinction between beliefs and knowledge might prove fruitful. In science education, we have traditionally considered knowledge, whereas "beliefs," although difficult to define (see Sayre, 1997), encompass affective commitments. A range of questions emerge, for example

> What is the relationship between beliefs and knowledge? Is belief a predictor of knowledge or visa versa? What are learners' beliefs about the nature of science? What are learners' beliefs about the future of the world? How do these beliefs relate to their knowledge? (see Chapter 6)

By considering learners' relationship with knowledge, future work might look to re-conceptualise learning and doing science as an activity through which one develops a sense of self (an intellectual, personal, and social identity). There is an extensive body of work tracing back to the 1920s, which has explored children's images of scientist. A more contemporary spin off to this might look to explore how children see themselves as a scientist. Learners might be asked to draw themselves as scientists, rather than the more traditional and depersonalised request to draw a picture of a scientist. This shift might offer a more fruitful consideration of self within science.

There is little doubt that psychoanalytical models of teaching and learning have much to offer future theorising (see Chapters 2, 3 and 11). In this realm, science education might be viewed as an object of a highly interactive and affective relationship between the self and subject. The process of science education might be redefined as an "object through which learners might therapeutically reclaim the sense of oneself as a moral, acting being" (see Appelbaum & Kaplan, 1998). This area is wide open for further empirical and theoretical consideration.

Teaching Science

There is little doubt that the teacher and the learning context create a variety of different emotional experiences which fundamentally influence/direct learning. And yet there is very little is known about the teaching and learning activities/environments which engage or disengage learning.

Osborne and colleagues (2003, p. 1073) suggest that science educators have much to learn from the growing body of literature on the study of emotional beliefs. An extensive psychological literature exists (largely external to science education), which has investigated the mediating and moderating role of individual and situated motivational beliefs. Rhee and colleagues in Chapter 7, bring our attention to four motivational constructs, self-efficacy, task value, interest, and achievement goals. Rather than considering motivation as a general trait, in this sociopsychological approach emphasis is given to the contextualised nature of motivation. Much future empirical work is needed to understand how pedagogical practices might draw on this scholarship. A cluster of questions seem to surface:

> How do classroom contextual factors influence motivational beliefs? How might science educators create experiences which successfully build learners' self efficacy, task value, interest and achievement goals? How do existing practices develop motivational beliefs?

Perhaps, as Dierking suggests in Chapter 9, formal education might learn much from informal education. After all, while students' attitudes towards science are found to decline in compulsory education after age 11 (Doherty & Dawe, 1988), a similar phenomenon is *not* always recorded in informal, out-of-school contexts. Future comparative research, this suggests, is much needed.

While the learning environment and the learner are inextricably linked, in many ways individual motivation and situational motivation might usefully be separated (see extended discussion in Hidi, 1990). As Hidi and Harackiewicz (2004) suggest, individual motivation is likely to develop more slowly, perhaps over a longer period of time, and could be longer lasting. In contrast, situational motivation is often evoked by something short lived and the reaction is episodic—transitory and immediate. As a consequence, individual motivation and situated motivation might be generated in different ways and research is still unclear about their interaction; situational motivation, for instance, might not be as closely linked to individual motivation as is often assumed. In the context of science education, while practical work might act as an efficacious classroom based motivational stimulus, it might not lead to an increased individual motivation for science education. Alternatively, although students' attitudes towards science might paint a gloomy picture they might only tangentially be linked to classroom-based motivation. This area is again wide open and many unresolved questions exist.

From socioconstructivist models of teaching and learning, the concept of emotional scaffolding is now emerging (see previous discussion). Roseik (2003) identifies two approaches to emotional scaffolding, implicit and explicit. Teachers in their study were found to directly or indirectly foster constructive emotions responses (or reduce unconstructive emotions) by explicitly bringing attention to emotions, or implicitly associating them with a familiar (or unfamiliar) context. Additional work is needed to consolidate these claims. This might explore an array of questions:

> How do science teachers' explicitly or implicitly scaffold learners' emotions? What metaphors and analogies do teachers choose to use in their emotional scaffolding? How do teachers draw on their knowledge of science and learners to select a particular approach? Whose emotional well-being is prioritised? Why?

Poststructuralist theories of emotion (e.g. Hochschild, 1983; Liston & Garrison, 2004; Lupton, 1998) and feminist theories (Boler, 2004) offer considerable promise as a means of extending existing discussions by understanding emotions in relationship to considerations of ideology, power, and culture. Such approaches often examine thinking and feeling through language and situated social practice (acts as part of a social performance). In this light, emotion is embedded within the sociopolitical forces that serve to construct educational reality and are embraced and legitimised by the dominant culture. Future work can learn much from the contemporary scholarship of Zembylas (2003, 2004). A spectrum of intriguing questions emerge, including

> How do educators and students perceive emotions in science education? How do teachers and students co-create the emotional culture of the science classroom? How does this culture legitimise particular behaviours while de-legitimising others? How might the hegemonic construction of emotion in the classroom serve to alienate disenfranchised learners?

Science Teacher's Emotions

There is an extensive literature charting teachers' knowledge, however much less is known about the role of teachers' emotions in teaching. Shifting the focus from knowledge to the combination of knowledge and affect raises future possibilities. For teaching practices to evolve we need to more fully understand why teachers' adopt particular strategies and approaches. How does the interaction of cognition and affect influence teachers' choice of teaching strategy? Does the relationship between emotion and teaching strategies change over time? Woolnough (1998), for instance, brings attention to the significance of confidence in subject knowledge. This raises additional questions; how does subject confidence relate to subject knowledge? How does confidence (or lack of confidence) manifest itself in teaching approaches? How might teachers effectively develop confidence in nonspecialist areas? How do teachers' emotions serve to influence the emotional climate of the classroom? Extending the established concept of pedagogical content knowledge (Shulman, 1987), future work might look to chart teachers' pedagogical content emotions—a combination of emotions, content, and pedagogy.

In Chapter 13, Shapiro provides a compelling account of teacher transformation. Understanding how teachers learn to teach has far reaching implications. Other questions might be investigated:

> How are beginning science teachers' emotions different from experienced teachers? Are they more fluctuating? How are beginning science teachers' emotions related to their classroom behaviours? How do emotion change with classroom experience? How are preservice teachers' "meta-emotions" (Sutton & Wheatley, 2003) related to their classroom experiences? And so on.

As the previous discussions suggest, including affect has the potential to stimulate a new body of research with fresh insights into the teaching and learning of science. In search of unity, the following discussions propose a multiplicity of approaches. While it is clearly impossible to cover all aspects of affect in a single volume, it is our hope that the richness and diversity of the collection serves to stimulate future research interests, agendas, and possibilities.

CONCLUDING THE INTRODUCTION

The introductory section comprises three chapters and I leave the remaining introductory comments to two senior colleagues. The first is an international scholar in science education; the second, a newcomer to this field, is a senior scholar in the field of psychoanalysis and education. In the following two chapters, they explore a common theme, psychoanalysis and its relationship with science.

In Chapter 2, discussion is lodged in the rarefied space between the "Two Cultures," consciously seeking to avoid the conflict of separation—the "Battle of the Books" as Jonathan Swift whimsically called it. Reiss's juxtaposition of evolutionary biology and psychoanalysis, nicely demonstrate how disparate research traditions might work in tandem to shed light on personal and social complexity, in this case human development. In this respect, I am often struck by how science education (a mix of scientists and social scientists) might offer an ideal venue, a melting pot, to fuse different research traditions. Reiss concludes by offering a vision of science education built on autonomous reasoning, self-determination, innovation and creativity, as well as social/political empowerment.

In Chapter 3, two psychoanalysts and one philosopher of science combine to underscore the inseparable nature of cognition and emotion. Borrowing Latour's metaphor of the black box, Alice Pitt elegantly lays down a gauntlet: a challenge for science educators to "open the box" and think about their "own histories of learning" as they help learners "make and remake their own biographies." In search of an open system pedagogy, Pitt's argument is derived from three vignettes: Freud's failure to find science, Winnicott's failure to connect mathematics with child development, and Stenger's insistence upon the failures of scientific. In making sense of these accounts, she accentuates the complexity of encounters with knowledge. Science educators, she suggests, can learn much from the stories of psychoanalysis, which might serve to both interpret and prepare learners' affective needs.

REFERENCES

Aikenhead, G. (1996). Science education: Border crossings into the subculture of science. *Studies in Science Education, 27*, 1–52.

Alsop, S. (1999). Understanding understanding: Modelling the public learning of radioactivity and radiation. *The Public Understanding of Science, 8*(1), 267–284.

Alsop, S. (2001a). Seeking emotional involvement in science education: Food chains and webs. *School Science Review, 83*(302), 63–68.

Alsop, S. (2001b). Living with and learning about radioactivity: A comparative conceptual study. *International Journal of Science Education, 23*(4), 263–281.

Alsop, S., & Watts, M. (2003a). Unweaving time and foodchains: Two classroom exercises in scientific and emotional literacy. *Canadian Journal of Science, Mathematics and Technology Education, 2*(4), 435–449.

Alsop, S., & Watts, M. (2003b). Science education and affect. *International Journal of Science Education, 25*(9), 1049–1079.

Alsop, S., Watts, M., & Hanson, J. (1998). Pupils perceptions of radiation and radioactivity: The wary meet the unsavoury. *School Science Review, 79*(289), 75–85.

Appelbaum, P., & Kaplan, R. (1998). An other mathematics: Object relations and the clinical interview. *Journal Curriculum Theorising,* Summer (14), 35–43.

Ashforth, B., & Humphrey, R. (1995). Emotion in the workplace: A reappraisal. *Human Relations, 48*, 97–125.

Baker, D. (2002). Editorial: Where is gender and equity in science education? *Journal of Research in Science Teaching, 39*(8), 659–663.

Boler, M. (1999). *Feeling power: Emotions and education*. London: Routledge.

Boler, M. (2004). Teaching for hope: The ethics of shattering world views. In D. Liston & J. Garrison (Eds.), *Teaching, learning and loving: Reclaiming passion in educational practice*. New York: RoutledgeFalmer.

Claxton, G. (1989). Cognition doesn't matter if you are scared, depressed and bored. In S. Adey, J. Bliss, J. Head, & M. Shayer (Eds.), *Adolescent development and school science*. London: Falmer Press.

Cornelius, R. (1996). *The science of emotion: Research and tradition in the psychology of emotion.* Englewood Cliffs, NJ: Prentice-Hall.

Csikszentmihalyi, M. (1988). Literacy and intrinsic motivation. In M. Csikszentmihalyi (Ed.), *Optimal experience: Psychological studies of flow in consciousness*. Cambridge: Cambridge University Press.

de Sousa, R. (1980). Emotions, education and time. *Metaphilosophy, 21*, 434–446.

Demasio, A. (2000). *Looking for Spinoza: Joy, sorrow and the feeling brain.* Toronto: Hardcourt.

Demasio, A. (2004). *Descartes' error: Emotion reason and the human brain*. New York: Quill.

Descartes, R. (1911). Passions de L'ame. In E. Haldane & G. Ross (Eds.), *The philosophical works of Descartes.* Cambridge: Cambridge University Press. (Original work published 1637)

Desmastes, S., Good, R., & Peebles, P. (1996). Patterns of conceptual change in evolution. *Journal of Research in Science Teaching, 33*(4), 407–431.
Dewey, J. (1931). *Experience and education in the later works of John Dewey, 1925–1953* (Vol 6). Carbondale and Edwardsville: Southern Illinois University Press.
Doherty, J., & Dawe, J. (1988). The relationship between developmental maturity and attitude toward science. *Educational Studies, 11*, 93–107.
Dole, J. A., & Sinatra, G. M. (1998). Reconceptualizing change in the cognitive construction of knowledge. *Educational Psychologist, 33*(2/3), 109–128.
Falk, J., & Dierking, L. (2000). *Learning from museums: Visitor experiences and the making of meaning.* Walnut Creek, CA: AltraMira.
Gilbert, J., Osborne, R., and Fensham, P. (1982). Children's science and its consequences for teaching. *Science Education, 66*(4), 623–633.
Giroux, H. (1992). Border crossings: Cultural workers and the politics of education. New York: Routledge.
Goldie, P. (2002). *The Emotions: A philosophical exploration*. Oxford: Oxford University Press.
Gregoire, M. (2003). Is it a challenge or a threat? A dual-process model of teachers' cognition and appraisal processes during conceptual change. *Educational Psychology Review, 15*(2), 147–179.
Harding, S. (1991). *Whose science? Whose knowledge? Thinking from women's lives*. Milton Keynes, United Kingdom: Open University Press.
Hargreaves, A. (1998). The emotional practice of teaching. *Teaching and Teacher Education, 14*(8), 835–852.
Head, J. (1989). The affective constraints on learning. In P. Adey, J. Bliss, J. Head, & M. Shayer (Eds.), *Adolescent development and school science*. London: Falmer Press.
Hidi, S. (1990). Interest and its contribution as a mental resource for learning. *Review of Education Research, 60*, 549–571.
Hidi, S., & Harackiewicz, J. (2004). Motivating the academically unmotivated: A critical issue for the 21st century. *Review of Educational Research, 70*(2), 151–179.
Hochschild, A. (1983). *The managed heart: Commercialisation of human feeling*. Berkeley: University of California Press.
Hodson, D. (1998). *Teaching and learning science: Towards a personalised approach*. Buckingham: Open University Press.
Kahle, J. (2004). Guest Editorial: Will girls be left behind? Gender differences and accountability. *Journal of Research in Science Teaching, 41*(10), 961–969.
Liston D. (2004). The allure of beauty and the pain of injustice in learning and teaching. In D. Liston & J. Garrison (Eds.), *Teaching learning and loving: Reclaiming passion in educational practice*. New York: RoutledgeFalmer.
Liston, D., & Garrison, J. (Eds.). (2004). *Teaching learning and loving: Reclaiming passion in educational practice*. New York: RoutledgeFalmer.
Lupton, D. (1998). *The emotional self: A socio-cultural explanation*. London: Sage.
Matthews, M. (1995). Science teaching: The role of history and philosophy of science. London: Routledge.
Osborne, J., Simon, S., and Collins, S. (2003). Attitudes towards science: A review of the literature and its implications. *International Journal of Science Education, 25*(9), 1049–1079.
Pintrich, P., Marx, R., & Boyle, R. (1993). Beyond cold conceptual change: The role of motivation beliefs and classroom contextual factors in the process of conceptual change. *Review of Educational Research, 63*(2), 168–199.
Rosiek, J. (2003). Emotional scaffolding: An exploration of the teacher knowledge at the intersection of the student emotion and the subject matter. *Journal of Teacher Education, 54*(5), 399–412.
Russell, B. (1947). *History of modern philosophy*. London: George Allan and Unwin.
Sayre, K. (1997). *Beliefs and Knowledge: Mapping the cognitive landscape*. Lanham, MD: Rowman & Littlefield.
Shulman, L. (1987). Knowledge and teaching: Foundations of the new reform. *Harvard Educational Review, 1*(57), 1–22.
Shweder, R., & LeVine, R. (1986). *Culture theory: Essays on mind, self and emotion*. Cambridge: Cambridge University Press.
Sjoberg, S. (2002). *Science for the children?* Report from the science and scientists-project. University of Oslo, Oslo, Norway: Department of Teacher Education and School Development.
Smith, E. (1991). A conceptual change model of learning science. In S. Glyn, R. Yeany, & N. Britton (Eds.), *The psychology of learning science* (pp. 43–64). Hilsdale, NJ: Earlbaum.

Songer, C., & Mintzes, J. (1994). Understanding cellular respiration: An analysis of conceptual change learning in college biology. *Journal of Research in Science Teaching, 31*(6), 627–637.
Strike, K., & Posner, G. (1992). A revisionist theory of conceptual change. In R. Duschl & A. Hamilton (Eds.), *Philosophy of science, cognitive science and educational theory and practice* (pp. 147–175). Albany, NY: Sunny Press.
Strike, K., Posner, G., Hewson, P., & Gertzog, W. (1982). Conceptual change and science teaching. *European Journal of Science Education, 4*(3), 231–240.
Sutton, R., & Wheatley, K. (2003). Teachers' emotions and teaching: A review of the literature and directions for future research. *Educational Psychology Review, 15*(4), 327–358.
Tobin, K., & McRobbie, C. (2001). Cultural myths as constraints to the enacted science curriculum. *Science Education, 80*(2), 223–241.
Vygotsky, L. S. (1978). *Mind in society: The development of higher psychological processes*. Cambridge, MA: Harvard University Press.
Vygotsky, L. S. (1986). *Thought and language*. Cambridge, MA: MIT Press.
White, R. (1994). Dimensions of content. In P. Fensham, R. Gunstone, & R. White (Eds.), *The content of science: A constructivist approach to its teaching and learning*. London: Falmer Press.
Woolnough, B. (1998). *Effective science teaching*. Buckingham: Open University Press.
Zembylas, M. (2002). Constructing genealogies of teachers' emotions in science teaching. *Journal of Research in Science Teaching, 39*, 79–103.
Zembylas, M. (2003). Emotions and teacher identity: A poststructural perspective. *Teachers and Teaching: Theory and Practice, 9*, 213–138.
Zembylas, M. (2004). Young children's emotional practices while engaged in long term science investigations. *Journal of Research in Science Teaching, 41*, 693–719.

CHAPTER 2

MICHAEL J. REISS

THE IMPORTANCE OF AFFECT IN SCIENCE EDUCATION

The most important goal, indeed, is not the quality of the scientific message or the pedagogy; the most important is whether the activities contribute to an actualization of the being.
(Perrier & Nsengiyumva, 2003, p. 1123)

INTRODUCTION

Science education has traditionally paid little attention to the emotions. Perhaps this is because some of the standard characteristics of science (even if subsequently contested by sociologists, feminists, and others)—summed up by Merton (1973) as open-mindedness, universalism, disinterest, and communalism—have widely, though implicitly and mistakenly, been taken as meaning that the objectivity of science somehow requires that science and the emotions operate in separate worlds; that emotions can safely be left to those who teach the arts and humanities.

And yet this is clearly nonsense for several reasons. First, any knowledge of the great scientists indicates the passion they frequently felt for their subject. Secondly, the disinterest of science is akin to the impartiality of judges in court and returning officers at elections. Such impartiality in no way requires judges not to feel anger, compassion, or other emotions when they adjudicate at a trial. Nor does it expect returning officers to eschew political views of their own and not to feel elation or disappointment depending on the result of an election; disinterest is required in the way they fulfil their duties not in what they feel. Thirdly, to leave the emotions to the arts and humanities is to remove them from the compass of science and this would be to narrow science inappropriately. Is not, for example, psychology, in so far as it studies the emotions, a scientific discipline?

If the distance between science and the emotions is smaller than generally supposed, how much more should this be the case for any "gap" between science education and the emotions?

So far I have only referred to emotions. What precisely is meant by affect? The meanings of words are in their uses. Some use "affect" and "emotions" as synonyms; some include emotions, feelings, moods, and attitudes within affect. While relatively little work until recently in science education has explicitly addressed affect, feelings, or the emotions (see Alsop & Watts, 2002; Hodson, 1998; Kahn & Kellert, 2002; Matthews, 2003; Watts & Alsop, 1997), there is a large literature on attitudes to school science.

Steve Alsop (ed.) Beyond Cartesian Dualism, 17–25.

Attitude studies raise substantial epistemological issues. As Judith Bennett (previously Ramsden) pointed out in a review, the word "attitude" is used, often uncritically, by researchers in a variety of ways (Ramsden, 1998). Frequently it is used interchangeably with such terms as "interest" and "motivation" or alongside such terms as "views" and "images." Bennett concluded "Where definitions, interpretations or explanations of terms are offered there appears to be a significant degree of overlap" (Ramsden, 1998, p. 127). A related issue is that "attitude" is not unidimensional: there are cognitive, emotional, and action-tendency components (Oppenheim, 1992).

Assuming that a suitable understanding (or understandings) of "attitudes" can be found, the question then arises as to how one attempts to determine or measure them in respect of science education. A number of techniques have been used, including subject choice at school and ethnographic tracking over time (Reiss, in press). However, the most frequent method is to use Likert-type scales. These have their uses—for example, they enable quantitative comparisons to be made between different groups (e.g. girls versus boys; pupils of different ages) and permit factor analysis (e.g. Parkinson et al., 1998)—but can suffer from problems to do with validity and reliability. In particular, Bennett notes "there are very few examples of studies where repeat measurement of attitude over time have been incorporated in the design phase of a research instrument . . . such an approach is based on the erroneous perception that attitudes are stable and unrelated to cognitive states" (Ramsden, 1998, p. 131).

Despite these caveats, certain conclusions can be drawn from the mass of attitude studies conducted over the last 30 years:

- During their school careers, most students lose interest in chemistry and physics
- The more technologically advanced a society (e.g. Japan, United States, Western Europe), the lower the interest of its students in school science
- School science is particularly criticised for its lack of "relevance"
- Girls' attitudes to school chemistry and physics are typically more negative than those of boys
- Good teaching, whether in school, college, or informal settings, can enhance student interest in science
- Correlations between student attitudes to science and their attainment are generally low to middling
- Students generally desire more opportunities in school science lessons for the exercise of personal autonomy

(abstracted from Osborne et al., 2003; Sjøberg, 2000)

All this is happening when there seems, though I am unaware of formal studies, to be no decrease, possibly an increase, in out-of-school interest in science as evidenced by the frequency of science programmes on TV and radio, science articles in newspapers and magazines, popular science books, the growth in science museums and science centres, and an increasing tendency for zoos, botanic gardens, and related organisations to portray themselves as sites of science education.

I want now to look at two perspectives on affect and learning—one from the viewpoint of evolutionary biology; the other from that of psychoanalysis. These two perspectives on the role of affect and its connections to learning, including learning in science, are very different but can, I believe, be seen to sit alongside each other, thus enriching one another.

AFFECT AND LEARNING FROM THE PERSPECTIVE OF EVOLUTIONARY BIOLOGY

Evolutionary biologists are increasingly writing about human behaviour and their books sell in large numbers (e.g. Jones, 1993; Ridley, 2003) even though they have had as yet only a negligible impact on school biology education. Evolutionary biology starts from the premise that humans are naturally self-interested. By "naturally" I mean that this is our typical inclination and has been since the dawn of time. For example, in the West some of us may feel a bit guilty at living more affluently than the great majority of humanity but we manage to more-or-less rationalise our self-interests away, perhaps by reasoning that we have children to look after or other dependent relatives—though precisely why our relatives are deserving of more than the average person is rarely explored, let alone explained. This is not the place to go into the fact that humans, more or less alone of all the 10 million plus species on this planet, have the capacity to go beyond self-interest. The point is simply that we have evolved to maximise our chances of surviving and reproducing—more strictly speaking, since the advent of sociobiology (Hamilton, 1964; Wilson, 1975), our chances and the weighted chances of our relatives.

To an evolutionary biologist, therefore, our emotions have adaptive significance. We either innately or through early experience and/or training feel disgust at such dangerous objects as feces and vomit (effective transmitters of human disease) and fear at such potentially dangerous objects as cliffs and snakes. Equally, we naturally feel affection for our immediate kin and, germanely for this chapter, we have, particularly in our youth, an enthusiasm for learning. We are, above all, an animal whose success depends on what we have succeeded in learning. The rare instances of children brought up with little or no human company for the first five or more years of their lives indicate how critical is our learning from one another (Newton, 2002). Without such human company we are truly disabled.

In our childhood we learn human language, we learn social customs, and we learn about the natural world. Although our capacity to remember certain new facts decreases as we age (though, interestingly, this may not be the case for certain things—e.g. new faces), we may develop other counterbalancing abilities, such as the ability to synthesise, to evaluate, to select from what we know, and to be aware of when to concentrate on remembering and when to relax. But whatever our age, we generally learn to like or love (to feel positive affect for) what is good for us. By and large, we feel this positive affect for our immediate family, for good food and clean water, for peace rather than war, for health rather than disease, and so on. By and large, exceptions and apparent exceptions prove the rule. The parent who endlessly grumbles at the behaviour of their child is distraught when that child goes missing; the fact that many of us like more sugar than is good for us is a relic of our evolutionary past when sugar was far less abundant and a reliable indicator that nutritious fruits were ripe for eating; although we are more pugnacious than many others species, in all human history it is actually only a minority of us that go to war and then typically only after we have been persuaded of the rightness of the cause.

AFFECT AND LEARNING FROM THE PERSPECTIVE OF PSYCHOANALYSIS

Psychoanalysts are interested in why we think, feel, and behave as we do. The discipline was founded by Sigmund Freud (1856–1939) who held that both instinctive and learned behaviours are important for us. Since its birth, psychoanalysis, as is true of almost every discipline, has split into various schools but all retain an especial interest in human development and in how the circumstances of our childhood can have lifelong repercussions. Further, while—to over-generalise somewhat for the sake of clarity—evolutionary biology is interested in whole species behaviours (practically all of us like clean water, good food, friendships, and so on), psychoanalysts are particularly interested in human differences. Why do people differ so much in how they take criticism? Why do people have such different sexual fantasies? Why are some people more creative than others? And so on.

Before we are born we are, in a sense, not a self. We are a part of our mothers and depend on them for everything. Of course, after birth we initially continue to depend on others for everything but we are no longer physically a part of another person. Birth, above all, is a detachment. With the arrival of our birth comes the end of the most physically intimate relationship we will ever know. One of the earliest tasks for each of us as we develop is to learn our boundaries—to learn the distinction between self and non-self (e.g. Rayner, 1986).

It is difficult to know what a young baby thinks and much of what it thinks almost certainly lacks self-consciousness (so that a young baby is aware of few if any of its thoughts) but in one sense a young baby thinks it is the universe. It has no understanding of the separation between self and other. As the baby, in her or his mind, begins to emerge as a distinct entity, anything that passes into it or out from it is of significance in this regard, especially if the baby has some control over the process. Urination and defaecation are both important for this reason. In each case a baby can feel itself filling up and then, as it lets go, that which was inside it leaves, and the accompanying tension dissipates. Skin too plays an important role in the development of self (Hinshelwood, 1991). Our skin does several things. For a start, it marks the physical boundary between inside us and outside us. And then it holds us together.

A baby does not suddenly learn its own boundaries. Such knowledge grows over the first 12 months or so. During this time the baby's brain is growing from only 25% of adult mass at birth (compared to 41% for a chimpanzee) to around 50% at its first birthday. Over this first year a baby undoubtedly learns a great deal about its environment though the fact that it cannot yet speak makes it difficult to discern precisely what it has learnt. By the time a child is 3 years old its brain is already 75% of adult mass. By now it will normally have a network of social relationships, know whether it is a boy or a girl, have a rapidly growing vocabulary, and be learning voraciously (Smith et al., 1998).

But what, if any, are the connections between what a young child feels (its affects) and what it learns about the world (its developing scientific understanding)? The psychoanalyst Melanie Klein (1882–1960) developed her play technique with young children as a direct analogy with the oral free association already used by psychoanalysts with adults. Each child had his or her own locker with small toys (such as cars, little figures, a few bricks, and a train), water, a basin, paper, scissors, glue, and so on. Klein watched and sometimes took part in the child's play, interpreting orally what she saw, to the child. Klein argued (and to her followers demonstrated) that even children as young as 6 months have a range of powerful emotions within them. They feel love and hate; they can be aggressive and want to destroy; they can be deeply anxious. By acknowledging such feelings Klein found that children who had got stuck in their development in some way were helped to move forward.

Active learning implies, in a sense, a dissatisfaction with one's existing knowledge. Klein employed the term "epistemophilia" (love of knowledge—i.e. natural curiosity) and argued that children have an urge to know certain things precisely because they appreciate that they are ignorant of them yet want to know about them. A classic example is sexual intercourse between the child's parents. Ignorance about sexual intercourse leads the child to view it as a violent activity. The combined parent figure (parents imagined in sexual intercourse) is therefore a threatening one and the child develops certain defences against this threat. These defences include striking back (in phantasy) against the parents. But as parents are more powerful than the young child, this is a dangerous strategy and a vicious circle can result in that attacks on the persecutors renders them potentially more harmful as the perceived threat of retaliation by them increases. This vicious circle indicates a paranoid state of hostility, with intense suspicion of any good figures. In some cases Klein concluded that children became trapped by these fears, leading to panic and night terrors. "Eventually she found, in a severely inhibited child, that these paranoid fears were so intense that they inhibited all activity, including the ability to create symbols" (Hinshelwood, 1991, p. 377).

Dreams were seen by Freud as symbolic alternatives to words; Klein showed that play was as symbolic as dreams. In some children, an inability to deal with symbols, manifested by an inability to play, was found by Klein to be associated with psychotic or near-psychotic behaviour. Such children had a particularly strong inhibition of curiosity. The more general lesson to be drawn is that when something goes wrong in the mental development of a child, the thirst for knowledge dries up. We can conclude that effective science teaching requires children to be capable of using symbols. To shift from the language of psychoanalysis to that of traditional science education, a child needs to be able to generate a mental model about a scientific phenomenon if it is to learn anything about it. If, for example, for whatever reason (development blockage, lack of motivation, inattention, tiredness, an inappropriate curriculum) a pupil in a science lesson cannot form a mental model of electrical circuits, failing to "see" the symbolic function of the lines that represent such circuits (as I remember I failed to see when I first saw circuit diagrams) then that pupil will be unable of understanding what (s)he is being taught however much that pupil can dutifully get light bulbs to light with correct degrees of brightness and dimness.

AN INTEREST IN LEARNING SCIENCE

Children learn from their infancy about living things in their immediate environment. In particular, they learn about animals, learning both to recognise different types of animals and what their basic names are (e.g. Rosch & Mervis, 1975). It has been argued that the concepts "animal" and "plant" are fundamental ontological categories, that is, categories used by children to organise their perceptions of the world in which they live (Keil, 1979). Certainly for most children animals form a significant part of the world around them, whether as wildlife, pets or zoomorphic toys. It is therefore unsurprising that names for familiar animals form a large part of the vocabulary of young children. Anglin (1977) reorganised the categories of words identified by Rins land (1946) in the early vocabularies of young children. He found that the largest of the resultant 22 semantic categories was that of everyday types of animals (36 words out of a total of 275). The next category was that of people, which contained 35 words.

All human learning takes places as a result of interactions between an evolved brain and a sensed environment and the human brain is perhaps the most impressive of all the products of evolution. Stephen Mithen (1996) has argued that the contemporary human mind can be envisaged as having evolved over time rather as a small mediaeval cathedral might have grown through the addition of chapels. Around a core "general intelligence" structure, chapels of "technical intelligence," "linguistic intelligence," "social intelligence," and "natural history intelligence" were added.

Natural history intelligence contains at least three subdomains of thought: that about animals, that about plants, and that about the geography of the landscape. For example, Mithen argues that Neanderthals must have possessed a natural history intelligence in order to hunt as they did:

> Neanderthals would have needed to get close to game for an effective use of their short thrusting spears. For this they had to understand animal behaviour and how to entice prey into disadvantaged situations: planning is essential to effective hunting, and knowledge of animal behaviour is essential to effective planning. Neanderthals could only have been successful at hunting large game if they had mastered the use of visual clues such as hoofprints and faeces, and possessed an intimate knowledge of the habits of their game. (Mithen, 1996, p. 129)

If Mithen is right, then our minds contain templates that have evolved to be more receptive to certain sorts of environmental stimuli than others. Children, on this theory, have an innate aptitude to learn about animals, plants and features of their environment in particular ways. That learning best takes place when what is being learnt is of personal meaning to the learner is hardly surprising, yet much of school education in general and science education in particular ignores this simple fact. When people want to learn about a particular aspect of science, they can learn a great deal about it (Irwin & Wynne, 1996; Layton et al., 1993).

Unfortunately, we persist in schools in telling children precisely what they ought to learn in science. Attempts to give pupils more autonomy over their learning, for example through the introduction of investigative work in science curricula, have had less success than might have been hoped for. This is very probable because even such "investigative" work is typically tightly controlled by teachers, both with respect to content and to procedure, and so ends up being less meaningful to pupils (cf. Roth, 1995). This is a great pity.

The lack of much authentic learning in school science is probably due to a number of reasons: a culture in science education in which teachers repeat how they were taught; pressures due to such things as resource constraints and large class sizes (which makes authentic practical work virtually impossible); and perhaps a fear in some teachers of what would happen if they let pupils have too much freedom. All this is at a time when there is a continued, even growing belief among philosophers of education that the best sort of education includes the development of autonomous reasoning and self-determination (Walker, 1999; Winch, 1999). School science education is only likely to succeed when pupils believe that the science they are being taught is of personal worth to themselves (Reiss, 2000).

Authentic science does not mean that you have to give pupils complete freedom. As every parent knows, you don't let your young children choose where the family will go on holiday—but on the holiday they must have a certain freedom of choice, even if only about whether to construct sand castles, run around or swim at a beach to which you have taken them. Indeed, as every artist knows, innovation and the exercise of true creativity require certain boundaries—and that is as true for Gehry and Liebeskind today as it was for Brunelleschi and Wren.

Good science teachers engage their students. There is no standard pattern to what a student finds engaging in science. For some the awesome age of the universe is engaging; for others it is understanding Newton's equations of motion, or watching a *Paramecium* under the microscope, or realising why 2 g of hydrogen react with 16 grams of oxygen, or participating in a field trip to the seashore, or making a rocket goes as high into the air as possible, or designing an electronic circuit that works as a burglar alarm. Engagements, though announced publicly, generally start as private affairs.

I know for myself as a child that one of the attractions of mathematics and the physical sciences was precisely their emotional coldness (as I then thought). The rationality of pure mathematics, of the Periodic Table and of calculations of the charge of an electron was a safer love than that of my fellow human beings. The physical world is a safe object (to use the psychoanalytical term) in which to invest one's affects when one is insecure. As Head (1996) points out, this characteristic is more typical of boys than of girls. Indeed, this is as predicted by psychoanalytical theory (Olivier, 1980/1989). As the young boy grows up he is aware that he and his mother are different—by virtue of his being male and her being female. This is not, of course, the case for young girls. The standard danger therefore for boys is that they separate from their mothers too early; for girls it is that they separate too late. Separating too early leads to difficulties in forming deep emotional attachments in which one can tolerate disappointments. Separating too late leads to difficulties in maintaining oneself in the face of a poor relationship—one is in danger of being emotionally suffocated and of subsuming one's own needs to the needs of the other.

Finally, we can note that a truly satisfying school science education would not only take account of the affects of individuals, it would help them to live in a social world. Here we are on more standard grounds. The argument that science education needs to tackle the impacts of science, needs humanising, and needs to consider issues of values raised particularly by such modern scientific technologies as biotechnology and nanotechnology, not only to motivate pupils but, more deeply, to enable pupils to live flourishing lives in an increasingly scientific–technological age is by now a well-rehearsed one (Cobern, 1998; Gräber & Bolte, 1997). However, we can go further. Science education has the potential to improve the world. It can "contribute to the advancement of democracy, and so improve the quality of human existence" (Longbottom, 1999, p. 4). It can enable students to participate actively in sociopolitical action (Roth &

Désautels, 2002). And it can help students strive for social justice (Reiss, 2003). A science education that took these aims on board would indeed be a science education to take issues of affect seriously and richly.

REFERENCES

Alsop, S., & Watts, M. (2002). Unweaving time and food-chains: Two classroom exercises in scientific and emotional literacy. *Canadian Journal of Science, Technology and Mathematics Education, 2*, 435–449.
Anglin, J. M. (1977). *Word, object, and conceptual development*. New York: W. W. Norton.
Cobern, W. W. (Ed.). (1998). *Socio-cultural perspectives on science education: an international dialogue*. Dordrecht: Kluwer.
Gräber, W., & Bolte, C. (Eds.). (1997). *Scientific literacy: an international symposium IPN 154,* Kiel: Institut für die Pädagogik der Naturwissenschaften an der Universitatät Kiel.
Hamilton, W. D. (1964). The genetical theory of social behavioiur, I, II. *Journal of Theoretical Behaviour, 7*, 1–52.
Head, J. (1996). Gender identity and cognitive style. In P. F. Murphy & C. V. Gipps (Eds.), *Equity in the classroom: Towards effective pedagogy for girls and boys* (pp. 59–69). London: Falmer Press; Paris: UNESCO.
Hinshelwood, R. D. (1991). *A dictionary of Kleinian thought* (2nd ed.). London: Free Association Books.
Hodson, D. (1998). *Teaching and learning science: Towards a personalized approach*. Buckingham: Open University Press.
Irwin, A., & Wynne, B. (Eds.). (1996). *Misunderstanding science: The public reconstruction of science and technology*. Cambridge: Cambridge University Press.
Jones, S. J. (1993). *The language of the genes: Biology, history and the evolutionary future.* London: HarperCollins.
Kahn, P. H. Jr., & Kellert, S. R. (Eds.). (2002). *Children and nature: Psychological, sociocultural, and evolutionary investigations,* Cambridge, MA: MIT Press.
Keil, F. C. (1979). *Semantic and conceptual development: An ontological perspective*. London: Harvard University Press.
Layton, D., Jenkins, E., Macgill, S., & Davey, A. (1993). *Inarticulate science? Perspectives on the public understanding of science and some implications for science education*. Nafferton: Studies in Education.
Longbottom, J. (1999). *Reconceptualising science education.* Conference paper, Second International Conference of the European Science Education Research Association, Kiel, 31 August–4 September.
Matthews, B. (2003). *Improving science and emotional development (The ISED Project): Emotional literacy, citizenship, science and equity* (2nd ed.). London: Gulbenkian Foundation.
Merton, R. K. (1973). *The sociology of science: Theoretical and empirical investigations*. Chicago: University of Chicago Press.
Mithen, S. (1996). *The prehistory of the mind: A search for the origins of art, religion and science*. London: Thames and Hudson.
Newton, M. (2002) *Savage girls and wild boys: A history of feral children*. London: Faber and Faber.
Olivier, C. (1989). *Jocasta's children: the imprint of the mother. Translated by George Craig*. London: Routledge. (Original work published 1980)
Oppenheim, A. N. (1992). *Questionnaire design, interviewing and attitude measurement*. London: Pinter.
Osborne, J., Simon, S., & Collins, S. (2003). Attitudes towards science: A review of the literature and its implications. *International Journal of Science Education, 25*, 1049–1079.
Parkinson, J., Hendley, D., Tanner, H., & Stables, A. (1998). Pupils' attitudes to science in key stage 3 of the National Curriculum: A study of pupils in South Wales. *Research in Science & Technological Education, 16*, 165–176.
Perrier F., & Nsengiyumva, J.-B. (2003). Active science as a contribution to the trauma recovery process: Preliminary indications with orphans from the 1994 genocide in Rwanda. *International Journal of Science Education, 25*, 1111–1128.
Ramsden, J. (1998). Mission impossible? Can anything be done about attitudes to science? *International Journal of Science Education, 20*, 125–137.
Rayner, E. (1986). *Human development: An introduction to the psychodynamics of growth, maturity and ageing* (3rd ed.). London: Unwin Hyman.

Reiss, M. J. (2000). *Understanding science lessons: Five years of science teaching*. Buckingham: Open University Press.

Reiss, M. J. (2003). Science education for social justice. In C. Vincent (Ed.), *Social justice, education and identity* (pp. 153–165). London: RoutledgeFalmer.

Reiss, M. J. (2004). Students' attitudes towards science: A long term perspective. *Canadian Journal of Science, Mathematics and Technology Education, 4*, 97–109.

Ridley, M. (2003). *Nature via nurture: Genes, experience and what makes us human*. London: Fourth Estate.

Rinsland, H. D. (1946). *A basic vocabulary of elementary school children*. New York: Macmillan.

Rosch, E., & Mervis, C. B. (1975). Family resemblances: Studies in the internal structures of categories. *Cognitive Psychology, 7*, 573–605.

Roth, W.-M. (1995). *Authentic school science: Knowing and learning in open-inquiry laboratories*. Dordrecht: Kluwer.

Roth, W.-M., & Désautels, J. (Eds.). (2002). *Science education as/for sociopolitical action*. New York: Peter Lang.

Smith, P. K., Cowie, H., & Blades, M. (1998). *Understanding children's development* (3rd ed.). Oxford: Blackwell.

Sjøberg, S. (2000). Science and scientists: The SAS-study. Cross-cultural evidence and perspectives on pupils' interests, experiences and perceptions: Background, development and selected results. *Acta Didactica, 1*, 4–74.

Walker, J. C. (1999). Self-determination as an educational aim. In R. Marples (Ed.), *The aims of education* (pp. 112–123). London: Routledge.

Watts, M., & Alsop, S. (1997). A feeling for learning: Modelling affective learning in school science. *The Curriculum Journal, 8*, 351–365.

Wilson, E. O. (1975). *Sociobiology: The new synthesis*. Cambridge, MA: Belknap Press of Harvard University Press.

Winch, C. (1999). Autonomy as an educational aim. In R. Marples (Ed.), *The aims of education* (pp. 74–84). London: Routledge.

CHAPTER 3

ALICE J. PITT

INCALCULABLE PRECISION: PSYCHOANALYSIS AND THE MEASURE OF EMOTION

For science educators, a solution to the riddle of emotion and the role it plays in learning is most likely to be sought behind one of two epistemological doors. Most familiar are various efforts to banish the effects of emotion on pure reason and pursuits conducted in its name. Educators, on this view, encounter emotional displays primarily as obstacles to the mastery of reason that is science: math phobia, squeamish tummies in the face of dissection, test anxiety, and so on. There are some more positive outcomes, too, found, for example, in the story of Barbara McClintock's "feeling for the organism" that so appeals to feminist critics of the "masculinity" of reason and science. More recently it has become possible to choose a second door, represented by cognitive psychologists and neuroscientists, who insist upon the role emotion plays in protecting us from danger and directing us toward choices and actions that ensure our survival. Educators interested in this view may seek ways to educate emotion right along with cognition. In this chapter, I present a view that complicates both of these approaches.

Working with two psychoanalysts and one philosopher of science, I explore the psychoanalytic view that emotion and cognition are, from the beginning, inseparable forces that make learning both possible and difficult. I begin with Freud (1905) and three questions posed by children that constitute, propel, and confound their search for truth about life in general and their lives in particular. Here early emotional life and its difficulties cast a shadow forward upon the child's relation to knowledge and her or his teachers. I then move to object relations theorist and paediatrician, D. W. Winnicott (1986) and his attempt to establish a congruence between the emotional development of the child and the basic structure of mathematical operations. In this example, we catch a glimpse of emotional difficulty as already residing within the curriculum and the knowledge it represents. Finally, I turn to Isabelle Stengers (1997), who considers Freud's failure to found a science and, by extension, Winnicott's failure to hold onto the threads of his argument, as representing the kinds of failure we can learn from. She returns the foundation laid by the child's three questions to the qualities of scientific knowledge itself. Stengers, as we shall see, makes a distinction between the pursuits of science and the practice of psychoanalysis that echoes Freud's views of the child as researcher and resonates with something Winnicott does not fully succeed in expressing about the work of teaching and learning.

These three examples—Freud's failure to found a science, Winnicott's failure to connect Mathematics with child development, and Stenger's insistence upon the failures of scientific reason—allow us new ways of thinking about ordinary learning as a

Steve Alsop (ed.) Beyond Cartesian Dualism, 27–36.

psychical and intersubjective event and ordinary teaching as already fully immersed in such learning though not, perhaps, in charge of it. Two metaphors, one drawn from Winnicott's understanding of the development of the baby from helplessness and dependence to relative autonomy, and one drawn from the history of science suggest that science educators need to think about their own histories of learning as they help their students make and remake their educational biographies. For Winnicott, *the incalculable precision of need* conceptualizes the end results of how teachers, students, and curriculum succeed and fail each other thus constituting teaching and learning as an open system made from surprises of revision, fantasy, and desire along with a measure of real progress and creativity. A second metaphor, *the black box*, was used by Bruno Latour (1988) to describe how one science invented its power. This metaphor views learning as a closed system yet to be perfected. Open systems carry their own risks, and closed systems are not without their appeal as science educators are well aware. If, however, as Freud concluded, Winnicott performs, and Stengers argues, learning, knowledge, and education are all made possible, ruined and repaired in relation to the very emotional unreason at the heart of reason, science education may also learn from its failures and take the risk of inventing itself as an open-system pedagogy.

WHAT CHILDREN WANT TO KNOW

Science as a method and epistemology and children as learners share at least one important quality: both embody an insatiable curiosity about the world, how it works, and what it means to be a part of it. Freud, who combined an interest in science with a curiosity about the formation of the human psyche, called children's original quest for knowledge their "infantile sexual researches" (Freud, 1905, p. 194). These researches, Paul Verhaeghe (2001) reminds us, revolve around three questions. The child is preoccupied by the *difference* between girls and boys, the *origin* of babies, and, finally, the *relationship* between the mother and the father. The nouns I have set into italics may well be the kernels of our infantile research that reverberate in our more adult pursuits. Is it not the case that the grand narratives of science are all about differences between things, the origin of things, and, finally, how the relationships between and among things work? We might say that poetry, too, and certainly philosophy, are similarly preoccupied, but they are not our concern here.

The analogy between the child's researches and the more mature work of the scientist is meant to be suggestive, not predictive or reductive. Yet it does hint at something beyond the familiar observation that traces of the child are visible in the adult. Verhaeghe describes the knowledge produced by the child as an amalgam of "primary fantasies, combining true with false and lack of knowledge into imaginary constructions" (2001, p. 36). The problem with these constructions is less that they are incorrect (they are bound to be) as that the answers they provide are never sufficient, and, so, "the questions persist" (p. 36). In the early days of his psychoanalytic explorations, Freud did believe that the cure for neuroses lay in providing accurate knowledge as a correction to faulty answers to the questions that perplex the child but may come to plague the adult. Soon, however, Freud became dissatisfied with his efforts.

Whether a patient claimed to accept or reject his interpretations, there was no reliable change in the patient in terms of a permanent alleviation of suffering. The truth-value of the knowledge Freud provided turned out to be less significant than how it *mattered* to the patient—it was a problem of making emotional significance. Freud's originality lay in coming to understand positive and negative responses as respectively giving too easily or withholding what the patient imagined the doctor wanted. Formulating this new insight led Freud to conceive of both kinds of responses as the effects of "the transference relationship by which the analyst is ascribed or refused the position of master" (p. 37). Psychoanalysis emerges at a point when didacticism cedes to the analysis of unresolved conflicts with beloved and feared authority figures as these are played out in the analytic setting.

This truncated sketch of a key component of Freud's development hints at how psychoanalysis makes sense of the risks and pleasures of learning. Difficulties arise where curriculum meets the singular residue of the child's theories and where the authority of teacher is reminiscent of the child's painful and pleasurable relations to those first authorities, the parents. Ordinary learning is hard to do because it combines a demand to relinquish fantasmatic theories with the possibility of falling in love with the substitutes of shared knowledge and their representatives (Pitt & Britzman, 2003). The original questions may persist, but we learn to press our infantile curiosity and its attendant anxieties into thinking about new questions that, we are promised, can be answered satisfactorily and satisfyingly. As a theory of learning, psychoanalysis takes into account what others cannot precisely because it conceptualizes learning as a psychical and intersubjective event that involves intimate relations among perception, emotion, and others. If learning is a psychical event, then so too is knowledge production, and scientific knowledge must be encountered, not only as a method and content for making sense of the world, but also as a passionate quest to repeat the pleasures and defend against the anxieties and failures of our original quests for knowledge.

From the motivated play between internal and external experiences, we can elaborate the problem of pedagogy in at least three ways. From the moment that Freud realized the futility of his efforts to tell the truth to his patients, he had to acknowledge that the intimate relations between inner and outer reality do not readily give up their power even when they are the source of distress. Psychoanalysis refers to this first difficulty as the problem of resistance. It is exacerbated by the fact that we can say little definitively about the causal forces at play in the generation and generativity of psychical reality. The problem of origins—where we come from—is lived in part as a difficulty in distinguishing what comes from the inside and what comes from the outside. Earlier I described this as imaginary constructions parading as truth. Finally, the difficulty of relinquishing imaginary constructions is a result of their attachment to our beloved and feared parental figures. As we have already seen, one of the central discoveries of psychoanalysis as a therapeutic practice concerns the way that the emotional tenor of past experiences and self-other relations are projected onto the present in the "transference." Past and present, in this formulation, mingle endlessly, each informing the other in a nonlinear passage of time.

BEFORE, BECOMING, AND LOSING "I"

I now turn to D. W. Winnicott's contribution to understanding learning and the pedagogical dilemmas faced by those who support learning (Winnicott, 1986). For this object relations theorist, the three questions animating the child's sexual researches can only be enjoyed and used as the foundation for learning if the primary helplessness of the infant has been successfully attenuated. In an address to the Association of Teachers of Mathematics in London in April of 1968, Winnicott presents an analogy between the child's emotional work of learning and the structure of Mathematics.

Winnicott, a well-known British psychoanalyst and paediatrician, was frequently called upon to speak to various groups, including learned societies, educators, and youth workers. On this occasion, just a few years before he died, Winnicott invented a whimsical title: *Sum, I am.* He wonders if his audience has caught the play between the Latin *sum*, meaning "I am" and the mathematical term for addition. Though well known for his playful spontaneity, the title is more than a charming and witty way for Winnicott to bridge the gap between his audience of Mathematics teachers and his own areas of expertise, which he describes as child psychiatry and "the theory of emotional development that belongs to psychoanalysis" (p. 55). Winnicott's title brings Mathematics and child development together. His talk focuses squarely on the range of emotional resonances a subject such as Mathematics has for children, and yet Winnicott refrains from providing the teachers in the audience with a clear idea about how they might use his ideas in their own practice. What is the teacher to make of an invitation to become curious about children's emotional life if such curiosity is not put in the service of improving Mathematics teaching or enhancing performance on the part of the learner? Winnicott does not answer this question in his essay, but a response can be crafted from his life-long concern with distinguishing between help that is useful and help that is experienced as coercion.

Winnicott does imagine that there exists some common ground between him and his audience when he presents his analogy between emotional development and Mathematics. He points out that "these matters that are the concern of the student of the human personality are also the concern of the mathematician" (1986, p. 56). How is this so? Winnicott says, "In a word, when I say that the central feature in human development is the arrival and secure maintenance of the stage of I AM, I know this is also a statement of the central fact of arithmetic, or (as one could say) of sums" (1986, p. 56). The relation between this central feature of development and the central fact of arithmetic that both revolve around the "one" sets the stage for a theory of learning that Winnicott sketches out in this essay:

> You will easily see what I am getting at: the idea that arithmetic starts with the concept of one, and that this derives and must derive in every developing child from the unit self, a state that represents an achievement of growth, a state indeed that may never be achieved. (1986, p. 58)

Securing the sense of I AM that permits mastering the operations involving whole numbers in such a way as to be part of making the self is less an operation of *separation* from the mother than it is the gradual realization of the fact of separateness. For Winnicott, it is important to remember that the mother has enjoyed an existence prior to and in the absence of the child, and he argues that her interest in re-establishing contact with her own existence is necessary to her baby's development of independence (Winnicott, 1971). Still, the idea that the mother has interests, pleasures, and needs that do not center upon or even include the baby is very difficult to experience and even more difficult to fashion as a site of security, pleasure, and growth. Different psychoanalytic schools theorize this problem in different ways. The achievement of the unit self is accompanied by the also gradual relinquishment of the fantasy of omnipotent control of the mother who, at least at the very early stages of life and in ideal conditions, anticipates or responds very quickly to the baby's needs—as quickly as is needed in order for this emergence of the sense of unit status not to come as too much of a shock altogether.

The formula, in the mathematical sense, for the delicate though ubiquitous operations leading to I AM assumes the infant's capacity to hold an image of the mother's existence and must factor in the level of frustration already tolerable to the infant, the presence or absence of prior breaches of toleration, and the baby's temperament. In the following scenario, Winnicott (1971) describes what happens when the baby feels distress by the mother's absence:

> The feeling of the mother's existence lasts x minutes. If the mother is away more than x minutes, then the image fades, and along with this the baby's capacity to use the symbol of the union ceases. The baby is distressed, but this distress is soon *mended* because the baby returns in $x + y$ minutes. In the $x + y$ minutes the baby has not become altered. But in $x + y + z$ minutes, the baby has become traumatized. (p. 95)

It is this odd paradox of incalculable precision that characterizes, for Winnicott, the affective realm of individual development. He recognizes three states along a continuum of development of the sense of I AM: (1) unit status has been achieved; (2) it has yet to be securely achieved; (3) the process of achieving it has been disrupted and growth is impaired. In his talk Winnicott gives an example of the second state when he describes a child who loves and mourns the death of a mouse. This child is doing some necessary emotional work in preparation for grasping mathematical concepts. This would be the insistence and significance of the minus sign. A child whose difficulties are of the order of the third state might have been a baby whose intellectual capabilities are *above* average. He describes what happens when a baby gets hungry and needs feeding. Much like the schema used describe the mother's return after an absence, Winnicott tells us that the feeding must arrive within a definite though unspecified amount of time or it becomes meaningless to the baby. By meaningless, Winnicott is signaling the use of the arrival of a feeding the baby as an answer to her or his call of need. If the feeding comes too late, it fills the tummy but does not serve this other important function. A baby who is well-endowed intellectually "soon gets to know from 'noises off' that a feed is being prepared" (1986, p. 58) and so predicts that the answer to the call is indeed on its way. Such a baby is less reliant on the mother's capacity for adapting to her baby's needs, but, according to Winnicott, this is not necessarily a positive factor for healthy development:

> Can you see from this that the intellect helps in the toleration of frustration? From this one can go on to see that a mother can exploit a baby's intellectual functions in order to get free from the tie that comes form the baby's dependence. All this is quite normal, but if you give the baby an intellectual equipment that is well above average, the baby and the mother may collude in exploitation of the intellect which becomes split off–split off, that is from the psyche of psychosomatic existence and living. (1986, p. 59)

Much of Winnicott's work involves the consequences of uneven emotional and intellectual development for individual happiness, the ability to form deep and lasting relationships with others, and the capacity for meaningful life and work. The split off intellect may not show up as a problem for accomplishing school tasks, but there is, in his view, a tremendous cost in terms of the child's diminished capacity for creative living and enjoyment. At this point in the essay, Winnicott turns to the continuum of development as it relates to Mathematics. In deceptively simple terms, Winnicott points out to his audience the familiarity they already have with the variations in children's achievement of primary wholeness:

> When you teach sums, you have to teach children as they come, and certainly you will recognize the three types:
> 1. Those who start easily with *one*.
> 2. Those who have not achieved unit status and for whom *one* means nothing.
> 3. Those who manipulate concepts and who are held back by banal considerations of pounds, shillings and pence. (1986, p. 61)

The first child can enjoy the manipulations of mathematical operations, but the second child cannot: "What I think you must not expect is that a child who has not reached unit status can enjoy bits and pieces. These are frightening to such a child and represent chaos" (1986, p. 61). Personal integration must precede the tackling of mathematical problems, and teachers do support such activity when they treat children's learning difficulties with calmness and patience. As for the third child, who does the work easily and perhaps even obsessively, Winnicott suggests that the child is incapable of making emotional significance from the skill. The analyst intuits what many a Mathematics educator has organized into coherent learning theory when he wonders why we don't ask these children to guess and to play with the creation of "ingenious methods."

Winnicott sums up his meditation on the relation between the achievement of unit status and activities involving basic operations such as addition and division by insisting that

> [t]eachers of all kinds do need to know when they are concerned *not with teaching their subject*, but with psychotherapy—that is, completing uncompleted tasks that represent parental failure or relative failure. The task I refer to here is one of giving ego support where it is needed. The opposite is to laugh at a child's failures, especially when these represent fear of forward movement and triumph. (1986, p. 63, emphasis in original)

There is something quite slippery about these assertions. Winnicott's earlier remarks suggested affinities between learning and failing to learn Mathematics and the child's emotional development that the teacher *witnesses* in her daily work. Here, there is an expectation, not only that the teacher modify her work in relation to the child's needs at a given moment, but that she be aware of how her engagement with a child is being used. Winnicott seems to have surprised himself for at this point his talk goes off in all directions.

Perhaps, he muses aloud, a child's learning difficulties can be attributed to the deleterious effects of an encounter with an unsympathetic teacher. Then again, we must understand that it is often the child herself whose insecurity or sensitivity makes her suspicious of even the most careful teacher. On a more positive note, Winnicott speculates that teachers gain enjoyment when, in fact, a student "*can catch on to the creative impulse*" (1986, p. 64) and the teacher's work is to spin that thread out as far as it can go. These musings are reminiscent of the calculation of the baby's capacity to survive the mother's absence. In this case, let us say that *x* stands for all the teacher provides, *y* stands for the curriculum, and *z* for what the learner needs at a given moment. There is something both necessary to and incalculable about the relations among these features of the pedagogical encounter that brings Winnicott to one last observation:

> Finally, why is it, I ask, that maths is the best example of a subject that can only be taught in continuity? If a stage has been left out, the rest is nonsense. Chicken pox, I think, accounts for many cases of mathematical breakdown..., and if you have time you coach the child over the bit that he or she missed.... (1986, p. 64)

Winnicott admits that what he has presented must seem like a muddle to his audience, but let us take seriously the idea, performed rather than presented as a logical argument, that there is something quite puzzling about components of education–student/teacher relations, the problem of learning, and the nature of knowledge. Winnicott's efforts to describe something of children's emotional development, far from clarifying teachers' understanding of these components and relations among them, seem only to have confounded the speaker. A large part of the problem, as Winnicott's jumble of speculations and observations attests, is the difficulty of distinguishing between the emotional difficulties the child brings to the tasks of learning and the more ordinary emotional difficulties inherent in learning itself. The good doctor seems to have arrived at this conclusion but failed to articulate its implications. Deborah Britzman (Britzman, 1998) has identified this puzzling dilemma as "difficult knowledge," and she offers a way out of the confusion between the Winnicott's two poles of instruction and psychotherapy:

> [T]he teacher is obligated to formulate theories of learning that can tolerate the human's capacity for its own extremes and its mistakes, resistance, belatedness, demands, and loss without creating more harm. (p. 19)

The work of becoming a teacher who can tolerate the strange indirection of learning, who is capable of enjoying the creative as well as the destructive moments of its ragged unfolding returns us to children's sexual researches and their preoccupation with origins, difference, and relationality. We have seen how one psychoanalyst approaches the risks and pleasures of learning and how one educational theorist makes from such an approach a pedagogical theory of curiosity and uncertainty. I now turn briefly to Stengers and her consideration of our three questions in relation to the development of science.

SCIENCE'S SUCCESS AND FREUD'S FAILURE: A REVERSAL

Following Bruno Latour (1988) and his work on Pasteur's isolation of bacteria, Stengers (1997) describes the successful invention of a particular science as a collective endeavor resulting in the closing of a black box. A black box, she writes, "establishes a relation between what enters it and what leaves it such that no one has, practically, the means to contest it" (1997, p. 86). She cites Latour's study of Pasteur's work as demonstrating that scientific arguments' singularity comes from the fact that they involve third parties: "Latour described Pasteur's work to have microorganisms recognized as witnesses that explain epidemics and to have himself recognized as their representative" (1997, p. 86). Freud set about to found a science, and the model he frequently aligned himself with was the development of chemical analysis. Chemistry, Stengers notes, was "during Freud's era . . . the queen of science" (1997, p. 94) and so an apt choice for an aspiring scientist of the human mind. Chemistry is also not like other sciences:

> Contrary to quantum mechanics or relativity, the reference to analytic chemistry cannot have bearing on a manner of description, a theoretical content, a lesson concerning the limit of our knowledges or their objectivity, but on an *operational technique*. (1997, p. 94, emphasis in original)

Lavoisier's technique invented the chemical fact, and Stengers describes three key qualities of this invention. First, his creation of a chemical fact brought something into existence that had not previously existed (for all practical purposes). He was able, therefore, "to overturn the relationship of forces between an individual and a tradition, that is, to make a *tabula rasa* of tradition" (p. 95). He accomplished this by controlling perfectly "the experimental 'scene' of the reaction system" (p. 95) and by purifying the elements brought together in reaction. These three terms—*tabula rasa* of tradition, control, and purification—are, for Stengers, the grounds for the black box which is 19th-century chemistry and, as well, the grounds for the black box which is Freud's desire for psychoanalysis. These terms also provide answers to the problems of origins, difference, and relation. The *tabula rasa* of tradition has no need to refer to anything outside of itself; it has invented itself and owes nothing. Control over the laboratory and what can happen there assumes that the relations between things can be predicted, made visible, and measured. Moreover, in Lavoisier's chemistry, these relations are manipulated, not by chance or prior history, but by the well-trained chemist. Finally, the need for purity in the elements eschews difference, maintains strict borders, and replicates endlessly its operations with little left over for the unfamiliar.

Now, the operational technique that Freud creates is the analysis of the transference, and he too strove for control over the analytic setting by virtue of the analyst's prior analysis and his neutrality. The object he created, for the purpose of curing neuroses, was the transference illness, that is, the artificial and temporary elaboration within the analytic setting of states of feeling that had their origins outside it. Stengers points out that, by the end of his career, Freud came to the unwanted conclusion that the power of his technique could, and indeed often was, resisted, thus obliterating its capacity to cure. Even more troubling was the acknowledgement that analysis of the transference does not eliminate, once and for all, the human tendency to gather up psychical conflict nor does it create an efficient barrier against new transference experiences. Finally, the analyst's neutrality, it is now widely acknowledged, is not so easily achieved. Freud's heirs are more likely to view the feeling states of the analyst as important tools for deeper understanding of the analysand's experience. Freud's failures, oddly enough, opened onto new horizons for understanding, not only the suffering and psychically vulnerable human being, but the human psyche itself as conflicted and subject to suffering. So, all is not lost as psychoanalytic theorists and practitioners struggle to understand, refine, and revise what it is about the psychoanalytic dialogue that does work and what precisely is happening when it fails to yield results. Stengers understands Freud's legacy this way:

> Psychoanalysts, as heirs not of a science that mimes hardness, such as behavioural psychology, but of what *could have been* a hard science, are among those whom we could hope would invent new ways of working together that are not centred on the possibility of judging, but that enable us to learn how to learn. (1997, p. 106, emphasis in original)

Freud was unsuccessful in slamming the lid down on a black box, but Stengers sees this failure as saying something about the nature of the human being and something about the contribution psychoanalysis makes to knowledge. Science educators, like other educators, will have to decide if they can become interested in the stories psychoanalysis tells about its own learning and ours. We may wish for techniques that will allow us to interpret and prepare for the child's emotional needs so that he or she becomes more focused on and successful at the tasks we present. But as Freud's failures attest and Winnicott's loss of control over his talk also demonstrates, we may have to satisfy ourselves with ongoing struggles to make sense of our own rhythms of learning and not learning, to listen closely to what our students are making of their lessons, and to talk about the surprising things we see and hear with our colleagues and friends. Like Winnicott, I think teachers already know this. Still, in a culture that prefers its black boxes firmly closed, it may not be what anyone can bear to hear. For science educators who are interested in making their own way with some of the ideas presented in this chapter, the work of Adam Phillips (1998, 1999) might serve as an inviting point of entry. In particular, his brief meditation on Freud and Darwin may well provide solace for thinking about these difficult matters:

> Both Darwin and Freud were fascinated by losses that could be survived—or even seen to be sources of inspiration—and by what survived, as evidence, of lives that had been lived. It was to these formative scenes of loss that they returned again and again in their writings. What could be made of what hadn't yet disappeared—the fossil record, or the half-remembered dream, species of birds or childhood memories—was their inspiration. It was the transience of things, the impermanence of natural phenomena, that fed them their best lines. Life was about what could be done with what was left, with what still happened to be there. (1999, pp. 116–117)

REFERENCES

Britzman, D. (1998). On becoming a "little sex researcher": Some comments on a polymorphously perverse curriculum. In *Lost subjects, contested objects: Toward a psychoanalytic inquiry of learning* (pp. 63–78). Albany: State University of New York Press.

Freud, S. (1905). Three essays on the theory of sexuality. In J. Strachey (Ed.), *The standard edition of the complete psychological works of Sigmund Freud.* (Edited and translated by James Strachey. In collaboration with Anna Freud. Assisted by Alix Strachey and Alan Tyson; Vol. 7, pp. 125–243). London: Hogarth Press and the Institute for Psychoanalysis.

Latour, B. (1988). *The pasturization of France* (A. Sheridan & J. Law, trans.). Cambridge: Harvard University Press.

Phillips, A. (1998). *The Beast in the nursery: On curiosity and other appetites.* New York: Pantheon Books.

Phillips, A. (1999). *Darwin's worms.* London: Faber and Faber.

Pitt, A., & Britzman, D. (2003). Speculations on qualities of difficult knowledge in teaching and learning: An experiment in psychoanalytic research. *QSE: International Journal for Qualitative Studies in Education, 16*(6), 755–776.

Stengers, I. (1997). Black boxes; or, is psychoanalysis a science? In *Power and invention: Situating science* (P. Bains, trans.; pp. 79–107). Minneapolis: University of Minnesota Press.
Verhaeghe, P. (2001). *Beyond gender: From subject to drive*. New York: The Other Press.
Winnicott, D. W. (1971). The location of cultural experience. In *Playing and reality* (pp. 95–103). London & New York: Routledge.
Winnicott, D. W. (1986). Sum, I am. In C. Winnicott, R. Shepherd, & M. Davis (Eds.), In *Home is where we start from: Essays by a psychoanalyst* (pp. 55–64). New York & London: W.W. Norton.

SECTION ONE: STUDENTS' ATTITUDES, HOPES, AND DISPOSITIONS

OVERVIEW

Attitudes, hopes, and dispositions are the focus of the three chapters in this section. In the following discussions, these affective referents are linked to an array of factors including conceptual change, student achievement, different curriculum content, instructional techniques, gender/equity, and environmental challenges. The overarching aim is to better understand the role of affect in cognition, the underpinning assumption being that these constructs significantly influence teaching and learning.

As far as the affective domain is concerned, scholarship in science education has gravitated towards the theoretical construct of "attitudes-towards-science." Even a cursory review of the field reveals the emphasis on this approach, which has a depth of work spanning several decades. Discussion of the psychological constructs, intentions, and dispositions is scarce, and at the very least the following discussions suggest that these emerging constructs have much to offer future work.

Throughout this section, an emphasis is placed on emotion as an individual psychological phenomenon. In different ways, the following discussions seek to better understand and categorise learners' responses to self-reporting tools. These tools, like most surveys, are decontextualised and detemporalised, and require respondents to reflect on previous experiences, usually in a predetermined response format.

I have arranged the chapters progressively; they start out by broadly reviewing existing work (attitudes-towards-science) and then become more specific, exploring particular aspects of science, the topics of evolution and photosynthesis, and environmental challenges. There is also a progression in student outcomes; earlier discussions concentrate on cognitive gains (student achievement), while the concluding chapter encompasses empowerment and action.

More specifically:

Moving from theory to the problems associated with measurement and concluding with a review of the field, Nieswandt, in Chapter 4, provides a review of students' attitudes-towards-science. The study of attitudes, she notes, is often beleaguered by a lack of clear definition because different researchers see and monitor attitudes in quite different ways. Too often the theoretical frameworks underpinning such studies are only partially articulated and as a consequence the normative data are lost against a broad diffuse background of cognitive, affective, and behavioural assumptions. Nieswandt, in contrast, starts her view from a theoretical perspective—Fishbein and Ajzen's theory of reasoned action. The chapter closes by underscoring the situated, dynamic nature of teaching and learning and points to motivational beliefs (including goal orientation) as a fruitful agenda for future research. Such research, the author suggests, has the

Steve Alsop (ed.) Beyond Cartesian Dualism, 39–40.

potential to venture towards a more sustained situated longitudinal investigation of teaching and learning in practice.

In Chapter 5, Schreiner and Sjoberg extend their internationally acclaimed qualitative work (see the Science and Scientist Project, Sjoberg, 2002). The chapter houses an exploration of action in a time of environmental challenge. Here the goal of science education is seen as providing the basis of empowerment and future environmental action. Within this context the affective constructs of hope, vision, control, and interest emerge as important. The empirical study documented is a survey involving 1204 Norwegian youth (aged 14–15), which contains clusters of questions exploring the respondents' vision of the future, engagement in environmental protection, and the individual and societal importance of environmental protection. Norwegian youth emerge as largely pragmatic, willing active but not overly concerned citizens who believe solutions to environmental challenges will be found in the future.

What makes controversial subjects difficult to teach? How do the emotions evoked by controversial science topics influence teaching and learning? This is the focus of Chapter 6. More specifically, Southerland and Sinarta's thought-provoking discussions examine the influence of controversy in the learning of human and animal evolution and compare this with learning the less controversial area of photosynthesis. The authors' evidence affect through the exploration of the psychological construct of *disposition* (learner's tendencies to use knowledge and beliefs to direct action toward particular learning goals). Two quantitative studies are used to compare three constructs in college students (education and biology majors): (1) understanding, (2) acceptance, and (3) dispositions relating to evolution and photosynthesis. The results indicate that the nature of the topic (degree of controversy) seems to influence acceptance and knowledge. The authors build on this conclusion, by highlighting recent models of conceptual change that have sought to incorporate affect.

REFERENCE

Sjoberg, S. (2002). *Science for the children?* Report from the science and scientists-project. University of Oslo, Oslo, Norway: Department of Teacher Education and School Development.

CHAPTER 4

MARTINA NIESWANDT

ATTITUDES TOWARD SCIENCE: A REVIEW OF THE FIELD

INTRODUCTION

Since Western societies vigorously promote the development of scientific literate citizens as an important task of schools, the lack of students' interest in science, or the under-representation of women and visible minorities in science and technology related professions are concerns that are discussed in political and educational circles. Research on attitudes toward science is one of the fields that addresses these concerns and has the potential to provide solutions for changing practice. Among educators and researchers alike, it is commonly assumed that students' attitudes in science influence their learning outcomes, their science course selections, and their future career choice (Koballa, 1988; Laforgia, 1988). Thus, changing attitudes should lead to changing behaviour. A look into various science education and educational psychology journals and other publications in this area reveals a fascinating potpourri of research on attitudes to science. Topics currently addressed in that literature include (i) students' attitudes toward schooling and different school subjects in comparison to science; and (ii) students' attitudes toward science as a discipline and a school subject. Other studies focus on the relations between attitudes toward science and (iii) different instructional strategies (e.g., hands-on, co-operative); (iv) areas of school science (e.g., environmental science, physical sciences); and (v) students' achievement. Some research projects look at (vi) the influence of teacher behaviour toward students' attitudes, or concentrate (vii) on the relationship between attitudes to science and variables external to the classroom such as age, gender, ethnicity, and grade level. It seems that research on attitudes to science and any other variable(s) is virtually endless. The following chapter will not even attempt to provide an overview of this huge and multifaceted research field. Instead, following an overview that describes attitudes, introduces a major theoretical model of attitudes, and describes the still unresolved issue of measuring attitudes, this chapter will critically summarize some of the current research on attitudes toward science. The focus of this literature review will be on studies that research attitudes toward science as the dependent variable. Finally, the chapter will close with further suggestions for research in the area of attitudes.

WHAT ARE ATTITUDES?

A definition of the term a*ttitudes* is quite difficult as an analysis of research studies discloses. In general, attitudes are defined as a predisposition to respond positively

Steve Alsop (ed.) Beyond Cartesian Dualism, 41–52.

or negatively to things, people, places, or ideas. Attitude contains affective, cognitive, and behavioural components (Simpson et al., 1994). Attitudes toward science refer to whether a person likes or dislikes science, or has "a positive or negative feeling about science" (Koballa & Crawley, 1985, p. 223). Other researchers (e.g., Simpson & Troost, 1982) identify attitudes as a subset of various categories including for example, science self-concept, attitudes toward the science teacher, the physical environment of the science class, and the science curriculum, and science anxiety. Oppenheim (1992) differentiates between different levels of attitudes that include cognitive and affective dimensions. Whereas "beliefs" and "images" are associated with the cognitive dimension illustrating what a person knows, for example the standard image of a scientist as a socially inept white man in a lab coat (Ryan & Aikenhead, 1992; Stein & McRobbie, 1997), the terms "values" and "personality" describe the affective dimension. They are seen as deeper levels of attitudes that are more stable and enduring than the cognitive components. As an overlapping term between cognition and affect Oppenheim (1992) proposes the category "views" that describes how a person responds to what she or he knows. Prior to Oppenheim, Shaw and Wright (1968, p. 13) defined attitudes with cognitive and affective components:

> Attitude is best viewed as a set of affective reactions towards the attitude object, derived from concepts of beliefs that the individual has concerning the object, and predisposing the individual to behave in certain manner towards the object.

This quote suggests a kind of two-stage process, first the cognitive, then the affective. However, I would propose that the cognitive and affective components influence each other in a kind of equilibrium. From this perspective, the affective and cognitive components of attitude mutually influence each other in an ongoing process, which in the end leads to behaviour.

In summary, this brief overview of definitions demonstrates that attitude is not a unidimensional term, but rather a multifaceted framework including affective, cognitive, and behavioural components. Thus, it is up to the individual researcher to define her/his understanding of attitude depending on the research objectives.[1]

A THEORETICAL MODEL OF ATTITUDE

Social psychologists have developed a variety of theoretical model of attitudes, in particular, because of the believed causal relation between attitudes and behaviour. In this chapter I will only concentrate on Fishbein and Ajzen's *theory of reasoned action* (Fishbein & Ajzen, 1975). (For further information on other theories, see Simpson et al., 1994, and Crawley & Koballa, 1994).

Fishbein and Ajzen's theory of reasoned action (Fishbein & Ajzen, 1975) is a social psychological model based on the assumption that the affective, cognitive, and behavioural aspects of attitude interact in a causal and unidirectional manner. Butler (1999) outlined the theory of reasoned action by taking research studies in science education into consideration that worked with the theory (see Figure 1). Based on the theory a person's behavioural intentions, which ultimately would lead to behaviour, can be predicted from her or his beliefs, evaluations, normative beliefs, and motivation to comply. *Behaviour* is defined as "an overt action under the volitional control

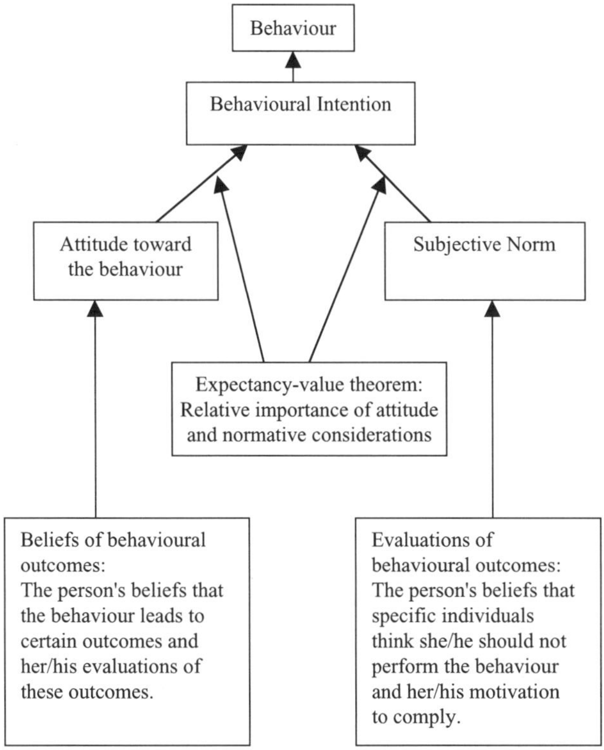

Figure 1. Factors determining a person's behaviour. Arrows indicate the direction of influence. (Source: Butler, 1999, p. 456)

and within the individual's capability" (Crawley & Coe, 1990, p. 463). Consequently, *behavioural intention* is characterised as a person's plan to act in a particular way and thus, closely related to behaviour. How do attitudes fit into this model? Attitudes are described as a person's evaluation of some specific behaviour, or as beliefs "that an individual holds about the consequences of engaging in a specific behaviour, a within-subjective effect or personal norm" (Crawley & Koballa, 1994, p. 37). This "general feeling of favourableness or unfavourableness toward some behaviour" (Crawley & Coe, 1990, p. 464) is called a*ttitude toward the behaviour*. The linkage between a decision for or against a particular behaviour or object and the consequences is labelled as *beliefs of behavioural outcome*. A person also holds beliefs "about the social pressure to engage or not engage in a behaviour" (Crawley & Coe, 1990, p. 464), which shape the *subjective norm*. A person anticipates "the likelihood of specific, personal consequences associated with engaging in a behaviour" (Crawley & Koballa, 1994, p. 37), or judges the consequences of performing a particular activity. In Figure 1 this is labelled as *evaluations of behavioural outcomes*. Finally, a person can decide whether she or he will perform the intended behaviour by selecting the alternative expected to lead to the most favourable outcome. This tendency is described in the *expectancy–value theorem* and can be interpreted as the link between attitudes and beliefs about the behavioural outcome with subjective norms and the evaluation of behavioural outcomes.

Ajzen (1985) extended the theory of reasoned action because of criticism of its limited applicability (Liska, 1984) by introducing the *theory of planned behaviour*. The theory considers a person's belief that she/he does neither have full control over her/his behavioural performance nor over the evaluation of "how easy or difficult performance or behaviour is likely to be" (Ajzen & Madden, 1986, p. 457). Internal factors like a person's skills or ability as well as external factors like the co-operation of others, or lack of resources, influence a person's behavioural performance. Labelled as *perceived behavioural control*, this factor has a direct influence on the formation of behavioural intention and neither attitudes nor subjective norms have any influence on it. However, control beliefs and the evaluation of controls are belief-based antecedents of perceived behavioural control. (For more detailed information, see Ajzen & Madden, 1986, and Crawley & Koballa, 1994).

Quite a few science education researchers have applied the theory of reasoned action in the field of science education research. For example, Koballa (1988) examined the determinants of female junior high-school students' intentions to enrol in elective physical science courses in high school. Results of this study show that the intentions were a function of both attitudes toward performing the behaviour and subjective norms. Crawley and Coe (1990) investigated middle school earth science students' intentions to enrol in an optional high-school science course. While external variables like sex, ethnicity, science ability, and general ability had no influence on the students' intention to enrol in the high-school science course, attitude toward enrolling and social support for doing so (subjective norm) were major predictors. The researchers also found that social support was more important for female students, whereas attitude was more important for male students. The influence of general ability and science ability on intentional course enrolment varied among students. For high-ability students the intentional course enrolment was a result of social support and for low-ability students of personal attitude.

HOW CAN WE ASSESS ATTITUDES TO SCIENCE?

The literature on attitudes toward science gives the impression that measuring attitudes is as multifaceted as defining this psychological construct. Closed item questionnaires (mostly based on Likert's Summative Rating Scale), open-ended questionnaires, and interviews are proposed as appropriate and equally valuable measurement techniques, although some attitude researchers are in favour of closed item questionnaire methods because of the time-consuming procedure of data analysis of the second and third technique (Krajkovich & Smith, 1982; Laforgia, 1988). Oppenheim (1992) recommends an inductive approach in developing an attitude inventory. First, interviews should be conducted to determine the nature and origins of the attitudes in the area of the research question (gathering and categorising free responses) that then in a second step, will be transferred into statements suitable for an attitude scale. (For further information on measuring of attitudes, see Aiken, 2002.)

Bennett and colleagues (2001) did a comprehensive critique on how attitudes are measured. They list seven problems with measuring attitudes ranging from the lack of a precise definition of attitude and related concepts, poor design of individual items or complete instruments, reliability and validity problems of the instruments, and lack of standardisation of instruments to inappropriate analysis and interpretation of data, lack of theoretical framework, and failure to relate the data collection tools with the theoretical framework. Based on these measurement problems the researchers developed an instrument that draws on the Views on Science–Technology–Society (VOSTS) large-scale study that was conducted in Canada by Aikenhead and Ryan (1992), applying the methodology in a new context (undergraduates' responses to the study of science). Furthermore, they followed Oppenheim's two-step approach (Oppenheim, 1992) as described above [(1) gathering and categorising free responses through interviews; (2) developing statements for the attitudinal scale], and included a validation of items involving participants who had not provided the free responses. This resulted in a subsequent revision of the instrument. The final items contain between four and six possible responses balancing "agree" and "disagree" responses. Bennett and colleagues (2001) outline major challenges they experienced throughout the instrument development. For example, due to the length of the instrument, they decided to administer the questionnaire over 1 week rather in a single class session. Finding an appropriate scoring system also proved difficult. However, the detailed description of their procedure for developing the instrument makes Bennett and colleagues' article an important contribution for research on attitudes toward science (Bennett et al., 2001). Although time consuming, their procedure provides an alternative to closed-ended item questionnaire approaches that claim unidimensionality while ignoring the heterogeneity of the population (Rennie & Parker, 1987).

A CRITICAL REVIEW OF CURRENT RESEARCH ON ATTITUDES

A thorough review of the literature confirms first impressions: research on attitudes toward science considers a vast array of topics. Concentrating on most recent published studies I will group them here by major themes related to attitudes (e.g., sex/gender issues, instructional strategies, and science programs) and after briefly describing the work within each theme, I will discuss their strengths and limitations. I will also refer to older studies that can be seen as "classics" in research on attitudes toward science.

The first group of studies that concentrates on how a student's sex (or gender—both are often used interchangeable) influences her/his attitudes toward science, and in relation to this her/his science course choice and achievement in science. Gardner (1975) and Schibeci (1984) reported that of all the variables that may influence attitudes toward science, sex has generally been shown to have the most consistent influence. Beginning as early as elementary school, girls show less interest in studying science than boys (Clark, 1972; Dawson, 2000; Kotte, 1992) and throughout middle and high school their attitudes toward science become less positive (American Association of University Women [AAUW], 1992, 1999; Greenfield, 1997; Rani, 2000; Sullins et al., 1995; Weinburgh, 2000). Cultural and social factors are seen as responsible for these trends, for example, differences in female and male students' out-of-school science experiences (e.g., Baker & Leary, 1995; Jones et al., 2000; Kahle & Lakes, 1983). The fact that gender seems to be related to variations in student interest in different areas of science may have a cultural base as well (Jones et al., 2000). In general, girls show more interest and experience in biological science, while boys are more oriented toward physical sciences (Jones et al., 2000). Reasons for these trends are seen in gender-biased classrooms in which teacher dedicate more attention to boys than to girls (Jones & Wheatley, 1990; Sadker & Sadker, 1994). Other studies argue that instructional strategies tend to be more closely oriented toward boys' interests (e.g., whole class activities) than toward girls' interests (such as small or co-operative group work) (e.g., Greenfield, 1997; Kahle & Meece, 1994). Thus, girls seem to work best when they are engaged in social interaction with peers, while boys prefer to work alone or less often with peers. Almost all of these studies finish by emphasizing the importance of gender-equitable instructional strategies (e.g. collaborative learning, hands-on activities; e.g., Greenfield, 1997) or the teacher's responsibility to present science equally for girls and boys (Jones et al., 2000).

However, research on how different teaching strategies affect students' attitudes and achievement in science do not always look at differences among girls and boys, but instead focus on the overall effect of these special strategies in comparison to a traditional classroom setting. For example, Soyibo and Evans (2002) show a positive effect of co-operative learning strategies on grade 9 students' understanding of human nutrition and attitudes toward biology in contrast to a control group that was taught using the lecture method. Gibson and Chase (2002) examined the long-term impact of a 2-week inquiry-based science camp on middle school students' attitudes toward science. They found that these students maintained a more positive attitude toward science and interest in science careers than a group of students who had applied to the camp but were not selected. In both studies as well in similar studies (e.g., Wong et al., 1997) the intervention was successful in changing attitudes, though, they do not mention whether these effects are seen among boys *and* girls. That is, it is possible that these overall improvements may have been generated through significant improvements for only one gender. In contrast, Freedman (2002) investigated the influence of a hands-on laboratory program on grade 9 students' attitudes toward science and their achievement levels in physical science and reported significant gender differences. Female students who had regular laboratory instruction scored significantly higher in achievement of science knowledge than girls without laboratory instruction (Freedman, 2002). Furthermore, girls account for the general positive effect of hands-on laboratory program on grade 9 students' attitudes. Other studies found similar gender effects on instructional strategies (e.g., Evans et al., 1995; Houtz, 1995). In general, these specially designed instructional approaches produce a more positive attitude change for girls than for boys. Moreover, these inquiry-based, hands-on, or co-operative learning approaches do not have a negative effect on male students.

Does merely changing teachers' instructional practice yield the long-term result of more females in science and science-related careers and in general, a more scientifically literate citizenry? Recent studies considering the link between instruction, attitudes, and these long-term effects are encouraging (e.g., Jarvis & Pell, 2002; Parker & Gerber, 2000; Stake & Mares, 2001). However, Jones and colleagues (2000) point out that despite our long-standing and well-established knowledge about attitude differences between female and male students, these differences still exist. Changing instructional strategies or developing whole programs that include these new strategies do not seem to have a convincing effect on girls' lack of interest in physics and boys' ignorance of biology. What other factors might affect attitudes? Some studies include analysis of the effect of ethnicity and socioeconomic status. The results are inconclusive with some reporting differences among ethnic groups' attitudes toward science (e.g., Catsambis, 1995; Chen, 2001; Weinburgh, 2000) and others, finding no differences (e.g., Rennie & Dunne, 1994; Wenner, 2003).

Other studies look at the relationship between adult influence and student attitudes, or between parents' competency beliefs and attitudes. For example, students' attitude scores are higher in classrooms in which they perceive greater leadership, helping, and understanding behaviours in their teachers (e.g., Bauer, 2002; Rickards & Fisher, 1999), or when students' parents are involved in their children's science experiences through library/museums visits and science activities (Rani & Kaplan, 1998). Parents' influence on the development of students' attitudes is found as an important factor (Andre et al., 1999; Steinberg, 1996) as well as their beliefs about their children's competencies in science. Andre and colleagues (1999) showed that parents of children in grade K-6 have gender-stereotyped beliefs about their children's competencies in science: science is perceived as more important for boys; they are expected to perform better; and jobs related to science and math are seen as more male dominated.

Finally, students' general cognitive abilities and their link to attitudes are explored (e.g., Shemesh, 1990; Sungur & Tekkaya, 2003). Sungur and Tekkaya (2003) claim that as students' reasoning ability increases, their attitudes toward biology become more positive. The study's design (reasoning ability and attitudes were measured at one time point) does not support such a causal direction. This study only can show a link but not a direction: attitudes could impact reasoning, or reasoning could impact attitudes or both. The authors' suggestion to use the learning cycle and inquiries as teaching strategies, because both foster scientific reasoning (Sungur & Tekkaya, 2003), seems nonetheless important and worthwhile following.

FUTURE RESEARCH ON ATTITUDES: WHAT STILL NEEDS TO BE DONE?

This brief and by no means comprehensive literature review[2] of mostly recent studies on attitudes to science demonstrates (1) that we have a dispersed knowledge base of various factors influencing attitudes toward science, and (2) that manipulating one of the measured variables or a combination of two does not necessarily influence students' attitudes toward the predicted and preferred outcome. In the context of Fishbein and Ajzen's theoretical model determining a person's behaviour (see above), these results suggest that students will less likely show a different behaviour. A change in behaviour depends in partial on changed attitudes. Thus, when students' attitudes toward science do not change, then it seems less likely that, e.g., more visible minorities will pursuit a career in science and technology related professions, or schools will develop scientific literate citizens. However, Fishbein and Ajzen's model include other components that are often ignored in research on attitudes: (1) the subjective norm, which can be explained as a person's beliefs about any social pressure to engage or not engage in a particular behaviour (Crawley & Koballa, 1994) and (2) internal factors like skills and external factors like co-operation of others. Transferred into the classroom these factors can be described as classroom environment, students' self-concept and self-efficacy, teaching methods (student-centered versus teacher-centered), multiple intelligences, motivation, etc. All these factors probably influence students' attitudes, however, the process by which these factors influence is far from apparent despite the vast amount of research in attitudes. Therefore, I propose that there is a need to view the classroom in its full complexity and dynamism. Classrooms as dynamic and fluid places are filled with individuals (students and teachers) who have a variety of reasons and motives to be in the particular class and to engage in learning. The implicit aim of a challenging science classroom is meaningful science learning: the ability to apply school science knowledge to everyday life situations such as evaluating medical treatment or making informed decisions regarding a bill on environmental issues. Thus, future studies should identify possible interdependence of three factors: affect–motivation–cognition with attitude as one of the affect variables. Integrating motivation as a separate entity results from the relationship between motivation, cognitive engagement, and conceptual development (Pintrich et al., 1993). Motivation is interpreted here through different *goal orientations*, which are drawn from goal theory—the goal of pleasing others, perceived instrumentality, learning goals, and performance goals (Ames, 1992; Ames & Archer, 1988; Dweck, 1986; Miller et al., 1996). A variety of studies (Greene & Miller, 1996; Miller et al., 1996; Montalvo et al., 1996) have show that students' degree of engagement in a task or activity and the strength of their beliefs about their own abilities are related to two goal orientations: The *desire to please others*, in particular the teacher, and *perceived instrumentality*, which can be described as an instrumental relationship between the learning task and a future goal, for example, attending university or entering a specific career. In contrast, *learning goals*, which focus on learning or mastering a task, and *performance goals* that concentrate on demonstrating one's competence and avoidance of one's incompetence (Ames, 1992; Dweck, 1986) are important to the acquisition of conceptual understanding (Alao & Guthrie, 1999; Ames, 1992; Blumenfeld, 1992; Dweck & Leggett, 1988). Students who endorse learning goals often show greater cognitive engagement and persistence in a task than those who indicate that they are oriented primarily toward performance goals (Ames & Archer, 1988; Greene & Miller, 1996; Nolen & Haladyna, 1990; Pintrich & DeGroot, 1990). Therefore, goal orientation is likely to have an influence on students' conceptual development of science concepts. Yet these studies stop short of exploring whether goal orientation ultimately influences students' conceptual development of science and how it influences students' affective component of learning. A theoretical model describing the influence of the different entities on each other could be ultimately the result of such studies.

Rather than imposing *a priori* a particular causal sequence, studies should be exploratory and inductive in nature, combining quantitative and qualitative approaches. A variety of methods (questionnaires, student interviews, and classroom observations) should be used to assess students' conceptual understanding in relation to the development of their interest, attitudes, and self-concept in science. Such a mixed method approach assumes that a particular attitude may express itself in different ways in different people (Oppenheim, 1992). Furthermore, the different levels of attitudes including the cognitive and affective dimensions as well as possible linkages among components of different attitudes (Oppenheim, 1992) are more likely to be brought to surface with in-depth interviews, which then can be the basis of meaningful questionnaire items.

Additionally, why restrict us to evaluating behavioural change only in retrospect? A specific science course selection can be the result of particular attitudes that were developed in prior science classes and influenced by different instructional strategies or a popular science teacher. Other factors such as social pressure or various instrumental goal orientations (e.g., aspiration to attend university, or make money in future job) may influence students' science course selection with or without impacts on their attitudes toward science in the long run. Studying the dynamic of affect–motivation–cognition longitudinally will result in a more comprehensive understanding of the various aspects that influence the development of attitudes in science. Furthermore, looking at the development of attitudes and intentional behaviour over time enables to assess whether attitudes and intentional behaviour are robust and deeply rooted, and thus, do not relapse back into the original attitudes after an intervention is finished. When it is our strong belief that changing attitudes toward science will lead to more females and visible minorities in science and science-related careers and to scientific literate citizens then we need to influence as many variables as possible in order to change students' attitudes toward science.

REFERENCES

Aiken, R. L. (2002). *Attitudes and related psychosocial constructs*. Thousands Oaks, CA: Sage.

Aikenhead, G., & Ryan, A. (1992). The development of a new instrument: Views on science–technology–society (VOSTS), *Science Education, 76*, 477–491.

Ajzen, I. (1985). From intention to actions: A theory of planned behavior. In J. Kuhl & J. Beckman (Eds.), *Action control: From cognition to behavior* (pp. 11–39). New York: Springer.

Ajzen, I., & Madden, T. J. (1986). Prediction of goal directed behavior: Attitude, intentions, and perceived behavioral control. *Journal of Experimental Social Psychology, 22*, 453–474.

Alao, S., & Guthrie, J. T. (1999). Predicting conceptual understanding with cognitive and motivational variables. *The Journal of Educational Research, 92*(4), 243–254.

American Association of University Women (AAUW). (1992). *Shortchanging girls, shortchanging America: A nationwide poll to assess self esteem, educational experiences, interests in math and science, and career aspirations of girls and boys ages 9–15*. New York: Marlowe.

American Association of University Women (AAUW). (1999). *Gender gaps. Where schools still fail our children*. New York: Marlowe.

Ames, C. (1992). Classrooms: Goals, structures, and student motivation. *Journal of Educational Psychology, 84*(3), 261–271.

Ames, C., & Archer, J. (1988). Achievement goals in the classroom: Students' learning strategies and motivation processes. *Journal of Educational Psychology, 80*(3), 260–267.

Andre, T., Whigham, M., Chambers, S., & Hendrickson, A. (1999). Competence beliefs, positive affect, and gender stereotypes of elementary students and their parents about science versus other school subjects. *Journal of Research in Science Teaching, 36*(7), 719–747.

Archer, J., & McDonald, M. (1991). Gender roles and school subjects in adolescent girls. *Educational Research, 33*(1), 55–64.

Baker, D., & Leary, R. (1995). Letting girls speak out about science. *Journal of Research in Science Teaching, 32*, 3–27.

Bauer, C.F. (2002). What students think: College students describe their high school chemistry class. *Science Teacher, 69*(1), 52–55.

Bennett, J., Rollnick, M., Green, G., & White, M. (2001). The development and use of an instrument to assess students' attitude to the study of chemistry. *International Journal of Science Education, 23*, 833–845.

Blumenfeld, P.C. (1992). Classroom learning and motivation: Clarifying and expanding goal theory. *Journal of Educational Psychology, 84*(3), 272–281.

Butler, M.B. (1999). Factors associated with students' intentions to engage in science learning activities. *Journal of Research in Science Teaching, 36*, 455–473.
Catsambis, S. (1995). Gender, race ethnicity, and science education in the middle grades. *Journal of Research in Science Teaching, 32*, 243–257.
Chen, H. (2001). Parents' attitudes and expectations regarding science education: Comparisons among American, Chinese-American, and Chinese families. *Adolescence, 36*(142), 305–313.
Clark, C. (1972). A determination of commonalities of science interests held by intermediate grade children in inner-city, suburban and rural schools. *Science Education, 56*, 125–136.
Crawley, F. E., & Coe, A. E. (1990). Determinations of middle school students' intention to enroll in a high school science course: An application of the theory of reasoned action. *Journal of Research in Science Teaching, 27*(5), 461–476.
Crawley. F. E., & Koballa, T. R. (1994). Attitude research in science education: Contemporary models and methods. *Science Education, 78*(1), 35–56.
Dawson, C. (2000). Upper primary boys' and girls' interest in science: Have they changed since 1980? *International Journal of Science Education, 22*(6), 557–570.
Dweck, C. S. (1986). Motivational processes affecting learning. *American Psychologist, 41*(10), 1040–1048.
Dweck, C. S., & Leggett, E. L. (1988). A socio-cognitive approach to motivation and personality. *Psychological Review, 95*(2), 256–273.
Evans, M. A., Whigham, M., & Wang, M. C. (1995). The effect of a role model project upon the attitudes of ninth-grade science students. *Journal or Research in Science Teaching, 32*(2), 195–204.
Fishbein, M., & Ajzen, I. (1975). *Belief, attitude, intention and behavior*. Reading, MA: Addison Wesley.
Freedman, M. P. (2002). The influence of laboratory instruction on science achievement and attitude toward science across gender differences. *Journal of Women and Minorities in Science and Engineering, 8*(2), 191–200.
Gardner, P. L. (1975). Attitudes to science: A review. *Studies in Science Education, 2*, 1–41.
Gibson, H. L., & Chase, C. (2002). Longitudinal impact of an inquiry-based science program on middle school students' attitudes toward science. *Science Education, 86*(5), 693–705.
Greene, B. A., & Miller, R. M. (1996). Influences on achievement: Goals, perceived ability, and cognitive engagement. *Contemporary Educational Psychology, 21*(2), 181–192.
Greenfield, T. A. (1997). Gender- and grade-level differences in science interest and participation. *Science Education, 81*, 259–276.
Houtz, L. E. (1995). Instructional strategy change and the attitude and achievement of seventh- and eighth-grade science students. *Journal of Research in Science Teaching, 32*(6), 629–648.
Jarvis, T., & Pell, A. (2002). Effect of the Challenger experience on elementary children's attitudes to science. *Journal of Research in Science Teaching, 39*(10), 979–1000.
Jones, M. G., Howe, A., & Rua, M. J. (2000). Gender differences in students' experiences, interests, and attitudes toward science and scientists. *Science Education, 84*, 180–192.
Jones, M. G., & Wheatley, J. (1990). Gender differences in teacher–student interactions in science classrooms. *Journal of Research in Science Teaching, 27*, 861–874.
Kahle, J. B., & Lakes, M. (1983). The myth of equality in science classrooms. *Journal of Research in Science Teaching, 20*, 131–140.
Kahle, J. B., & Meece, J. (1994). Research on gender issues in the classroom. In D. L. Gabel (Ed.), *Handbook of research on science teaching and learning* (pp. 542–557). New York: Macmillan.
Koballa, T. R. Jr. (1988). Attitude and related concepts in science education. *Science Education, 72*, 115–126.
Koballa, T. R. Jr. (1995). Children's attitudes toward learning science. In S. M. Glynn & R. Duit (Eds.), *Learning science in the schools: Research reforming practice* (pp. 59–84). Mahwah, NJ: Lawrence Erlbaum.
Koballa, T. R. Jr., & Crawley, F. E. (1985). The influence of attitude on science teaching and learning. *School Science and Mathematics, 85*, 222–232.
Kotte, D. (1992). *Gender differences in science achievement in 10 countries*. Frankfurt: Peter Lang.
Krajkovich, J., & Smith, J. (1982). The development of the Image of Science and Scientists Scale. *Journal of Research in Science Teaching, 19*, 39–44.
Laforgia, J. (1988). The affective domain related to science education and its evaluation. *Science Education, 72*, 407–421.

Liska, A. E. (1984). A critical examination of the causal structure of the Fishbein/Ajzen attitude–behavior model. *Social Psychology Quarterly, 47*(1), 61–74.

Marlow, E. (2002). Assessing teacher attitudes in teaching science. *Journal of Instructional Psychology, 29*(1), 25–28.

Miller, R. B., Greene, B. A., Montalvo, G. P., Ravindran, B., & Nichols, J. D. (1996). Engagement in academic work: The role learning goals, future consequences, pleasing others, and perceived ability. *Contemporary Educational Psychology, 21*(4), 388–422.

Montalvo, G. P., Krows, J., & Miller, R. B. (1996). *The effects of liking versus disliking the teacher on student motivation.* Roundtable presented at the Annual Meeting of the American Educational Research Association, New York, NY.

Nolen, S. B., & Haladyna, T. M. (1990). Motivation and studying in high school science. *Journal of Research in Science Teaching, 27*(2), 115–126.

Oppenheim, A. N. (1992). *Questionnaire design, interviewing and attitude measurement.* London: Pinter.

Parker, V., & Gerber, B. (2000). Effects of a science intervention program on middle-grade student achievement and attitudes. *School Science and Mathematics, 100*(5), 236–242.

Pintrich, P. R., & DeGroot, E. V. (1990). Motivational and self-regulated learning components of classroom academic performance. *Journal of Educational Psychology, 82*, 33–40.

Pintrich, P. R., Marx, R.W., & Boyle, R. A. (1993). Beyond cold conceptual change: The role of motivational beliefs and classroom contextual factors in the process of change. *Review of Educational Research, 63*, 167–199.

Rani, G. (2000). Measuring change in students' attitudes toward science over time: An application of latent variable growth modeling. *Journal of Science Education and Technology, 9*(3), 213–225.

Rani, G., & Kaplan, D. (1998). A structural model of parent and teacher influences on science attitudes of eighth graders: Evidence from NELS: 88, *International Science Education, 82*, 93–109.

Rennie, L. J., & Dunne, M. (1994). Gender, ethnicity, and students' perceptions about science and science-related careers in Fiji. *Science Education, 78*(3), 285–300.

Rennie, L. J., & Parker, L. H. (1987). Scale dimensionality and population heterogeneity: Potential problems in the interpretation of attitude data. *Journal of Research in Science Teaching, 24*, 567–577.

Rickards, T., & Fisher, D. (1999). Teacher–student classroom interactions among science students of different sex and cultural background. *Research in Science Education, 29*(4), 445–455.

Ryan, A. G., & Aikenhead, G. S. (1992). Students' preconceptions about the epistemology of science. *Science Education, 76*(6), 559–580.

Sadker, M., & Sadker, D. (1994). *Failing at fairness: How America's schools cheat girls.* New York: Scribner's.

Schibeci, R. A. (1984). Attitudes to science: An update. *Studies in Science Education, 11*, 26–59.

Shaw, M. E., & Wright, J. M. (1968). *Scales for the measurement of attitude.* New York: McGraw-Hill.

Shemesh, M. (1990). Gender-related differences in reasoning skills and learning interests of junior high school students. *Journal of Research in Science Teaching, 27*, 27–34.

Simpson, R. D., Koballa, T. R. Jr., Oliver, J. S., & Crawley, F. E. (1994). Research on the affective dimension of science learning. In D. L. Gabel (Ed.), *Handbook of research on science teaching and learning* (pp. 211–234). New York: Macmillan.

Simpson, R. D., & Troost, K. M. (1982). Influences on commitment to learning of science among adolescent students. *Science Education, 66*, 763–781.

Soyibo, K., & Evans, H. G. (2002). Effects of co-operative learning strategy on ninth-graders' understanding of human nutrition. *Australian Science Teachers' Journal, 48*(2), 32–35.

Stake, J. E., & Mares, K. R. (2001). Science enrichment programs for gifted high school girls and boys: Predictors of program impact on science confidence and motivation. *Journal of Research in Science Teaching, 38*(10), 1065–1088.

Stein, S. J., & McRobbie, C. J. (1997). Students' conceptions of science across years of schooling. *Research in Science Education, 27*(4), 611–628.

Steinberg, L. (1996). *Adolescence.* New York: McGraw-Hill.

Sullins, E., Hernandez, D., Fuller, C., & Tashiro, J. (1995). Predicting who will major in a science discipline: Expectancy–value theory as part of an ecological model for studying academic communities. *Journal of Research in Science Teaching, 32*, 99–119.

Sungur, S., & Tekkayz, C. (2003). Students' achievement in human circulatory system unit: The effect of reasoning ability and gender. *Journal of Science Education and Technology, 12*(1), 59–64.

Taber, K. S. (1991). Gender differences in science preferences on starting secondary school. *Research in Science & Technology Education, 9*(2), 245–251.
Weinburgh, M. H. (2000). *Gender, ethnicity, and grade level as predictors of middle school students' attitudes toward science*. ERIC, ED442662.
Wenner, G. (2003). Comparing poor, minority elementary students' interest and background in science with that of their white, affluent peers. *Urban Education, 38*(2), 153–172.
Wong, A. F. L., Young, D. J., & Fraser, B. J. (1997). A multilevel analysis of learning environments and student attitudes. *Educational Psychology, 17*(4), 449–468.
Young, D. J., & Fraser, B. J. (1994). Gender differences in science achievement: Do school effects make a difference? *Journal of Research in Science Teaching, 31*, 857–871.

NOTES

[1] Attitudes toward science should not be mistaken with *scientific attitudes* that, as Koballa and Crawley suggest (1985), are better labelled as *scientific attributes*. Scientific attitudes embody "the characteristics or attributes of scientists that are considered desirable in students"; Koballa, 1995, p. 62.

[2] For example topics like the relationship between attitudes and various subject matter topics (e.g., Taber, 1991), between attitudes and sex-role and career stereotyping (e.g., Archer & McDonald, 1991), or between teachers attitudes toward science and students' attitudes (e.g., Marlow, 2002) were not discussed at all.

CHAPTER 5

CAMILLA SCHREINER AND SVEIN SJØBERG

EMPOWERED FOR ACTION? HOW DO YOUNG PEOPLE RELATE TO ENVIRONMENTAL CHALLENGES?

INTRODUCTION

Creating environmentally active citizens is crucial for future environmental development. Through the slogan "Science education for action," Jenkins (1994), for example, addresses (among other subjects) the environmental protection issue and calls for "integration of knowledge with action." The background for this chapter is that we subscribe to the notion of responsible and successful action as a prime goal of science and environmental education, and that we wish to bring into light some essential conditions for purposeful action. We see environmental *empowerment* as a prerequisite to action. On the assumption that successful environmental action requires environmental empowerment, we argue that empowering young people to deal responsibly with environmental issues should be a principal concern of education. Empowerment may be described as encouragement for action and belief in one's possibility and ability to influence one's surroundings. It is important to understand the attitudes, beliefs, and prejudices that might prevent individuals from recognising and using their possibilities to act.

We are aware of the debate among educators, environmentalists, and scientists concerning purposes of schooling in general and science education in particular. One view is that education should equip the students with knowledge and skills, but that it is up to the students themselves to decide *how* to apply these competences; otherwise education risks becoming indoctrination. While we acknowledge that this view makes an important point, we argue that some values, e.g. values of democracy, peace, equity, human rights, and environmental protection, are universal. The German philosopher and educator Wolfgang Klafki (2001) sees the mission of education as inseparable from the challenges facing a society, and he characterises the environmental issue as one of the four key problems facing our time. We concur with this view and will not categorically disallow science education to act as agent of influence on students' environmental value preferences.

On this basis, we see science education as having a key role in preparing young people to cope and deal responsibly with the emerging environmental challenges. Teaching must be based on knowledge of students' attitudes to the environmental protection issue. Research in science education has taught us a lot about students' conceptual understandings (and "misconceptions" or "alternative conceptions") of

Steve Alsop (ed.) Beyond Cartesian Dualism, 53–68.

science concepts, but less about their attitudes, priorities, and decision making regarding environmental matters. In this chapter, we are aiming at deepening our understanding of challenges facing us as science educators in our endeavour to develop students empowered for environmental action by presenting findings from analysis of survey data. The ideological and theoretical perspectives on which this chapter is based, is more thoroughly described by Schreiner and colleagues (in press). The authors assume that in order to be empowered to meet the environmental challenges, a person must

- have hope and visions for the future
- have a general feeling that she or he can influence the future of the world and be motivated for action towards environmental issues
- think that environmental protection is important for society
- be interested and engaged in the issue

She or he must also have sufficient knowledge about the science of the environment, about possible adequate actions in terms of personal lifestyle, technical solutions, and political measures and about possible channels of influence through politics, organizations, etc. Here we will concentrate on empirical findings addressing the bullet issues above and take in some few pieces from the above referred article on perspectives from literature (Schreiner et al., in press).

Our data are collected through the ROSE survey. ROSE, The Relevance of Science Education, is an international comparative research project, which aims to shed light on affective factors linked to the learning of science and technology (S&T). The target population are students towards the end of secondary school (age 15). The research instrument is a questionnaire mostly consisting of closed questions that offer the respondents fixed alternative responses. The respondents give their answers by choosing the alternative appropriate to their view. Among other issues, the questionnaire addresses their interests in learning different S&T topics, their experience with and views on school science, their views and attitudes to science and scientists in society, their future hopes, priorities, and aspirations, and their feeling of empowerment with regards to environmental challenges. This latter point will be the focus of this chapter. The rationale behind the project, including the questionnaire development, theoretical background, procedures for data collection, etc., is described in Schreiner and Sjøberg (2004). This report, as well as other information on the ROSE project, can be found at http://www.ils.uio.no/forskning/rose/

International comparisons may give important insight into the diversity and similarities in youth's views in different cultures. On the other hand, national analysis facilitates a deeper view of the material, as relationships between variables, reliability, validity, etc. are properties of the data rather than of the instrument. Since data collected with one instrument may result in different indices in different cultures, cross-cultural analysis calls for close inspection of each national sample. Including international comparisons would bring this text beyond the specified volume limit. Therefore, in this publication we will only report results from the Norwegian material. However, our preliminary experiences with the ROSE data suggest that results from the Norwegian sample might follow the same pattern as other North-Western European countries (e.g. countries in the United Kingdom and Scandinavia). This profile often contrasts with the profile of less economically developed countries in Africa, Asia, and South America.

METHOD

The survey was conducted in Norway in November and December 2002. Fifty-eight schools and one class at each school were randomly sampled. This gave a total sample of 1204 Norwegian respondents, broadly representative of the population of all Norwegian 10th grade students (students aged 14–15). Further details on how the ROSE survey was organized are given in Schreiner (2004).

Most questionnaire items follow the same basic structure: A statement is presented and the students are requested to give their response by choosing the appropriate box in a 4-point Likert scale. The four response categories go from "small" to "large": Disagree–Agree, Not interested–Very interested, Not important–Very important, etc. A response in the first category (Disagree, Not interested, Not important, etc.) is coded 1, in the second 2, etc. A response in the last category (Agree, Very interested, Very important, etc.) is consequently coded 4.

It is common practice to code the scale by assigning numerical values to the response categories, and to regard Likert scales as quasi-interval scales (Ary et al., 1996). However, when handling the coded data as values in an interval scale, we presuppose that the distance from category 1 to category 2 is identical to the distance from category 2 to category 3, etc. In methodology literature, issues like these are debated, but is seems to be a wide acceptance to use Likert scales as we are indicating here.

In order to overcome the amount of data, condense the characteristics of the questionnaire items, achieve more reliable and valid data, and to enable us to lift the discussion up from single-item responses to a more general level (Hellevik, 2002), some questionnaire items have been grouped into clusters, checked for unidimensionality and internal consistency and then, if found reasonable, merged into composite variables. The composite variables applied in this chapter are developed from the average scores of the items constituting the variables. The number of items in each composite variable differs. Adequate interpretation of the common factor underlying one group of items is crucial for valid understanding of findings based on scores in a composite variable.

Before merged into the various composite variables, the items were divided into clusters by drawing on a combination of exploratory factor analysis, the original intention of the items and reliability testing with Cronbach's alpha. Cronbach's alpha is a measure of internal consistency within a group of items, based on item covariances. The maximum value of the coefficient is 1. Because cognitive skills tend to be more stable than affective features, cognitive measures often report alphas in the high .80s or low .90s, while .70 is one widely accepted cut-off point for alpha in affective instruments (Gable & Wolf, 1993; Nunnally, 1978). From a Cronbach's alpha of .70, we can interpret that 30% of the composite variable variance is error variance, while 70% can be considered as true variance.

The output from a statistical significance test is the probability that we, based on findings from the sample, are claiming a false difference or relation between variables for the whole population. The significance test takes, among other parameters, into account the sample size. In a study like ROSE, with a large sample size, most differences and relations between variables are found to be statistically significant at the 1% level. When our findings are not statistically significant (i.e. higher than 1%), the parameters always have vanishing effect sizes (i.e. standardised differences). Unless something else is given, all differences and relations presented in this chapter are statistically significant at the 0.01 level.

RESULTS AND SOME PERSPECTIVES FROM LITERATURE

In this section we will present variables and scores both for composite variables and for some single items that have not been employed in any composite variable. Rather than having one theoretical section with the framework for understanding our findings, we will involve perspectives from literature as and when it is relevant. Gender differences will be reported in cases where we consider the magnitude of the differences to be of educational interest.

Do Young People have Hopes and Visions for the Future?

Beliefs about what the future will bring contributes to the meanings one gives to the present (Bell, 1997). People's images of the future influences their actions in the present, as people either try to adapt to what they see coming, or try to act in a way that creates the future they wish for. Hope encourages action (Eckersley, 2002). Future images are influenced by the background, experiences, knowledge, etc. of each individual and by social and cultural factors such as mass media, public discussions, and the Zeitgeist of the era and the society. By knowing the youth's images of the future, we can better understand their present motivation, choices, and actions. The images students hold of the future are consequently of interest to science and environmental educators (Hicks, 1996; Lloyd & Wallace, 2004; Palmer, 1998).

In 1974, Alvin Toffler disclosed a discrepancy between the personal and the global images U.S. youth held of the future (Toffler, 1974). Since then, numerous studies of youth in Western societies have confirmed his finding of personal optimism and global pessimism—the further the images go from the personal level, the darker and more hopeless they get: Young people's images of their personal futures are optimistic and full of hope. With focus on education, nuclear family, occupation, and leisure, they feel able to design and create their own good and happy personal future. When it comes to the local and national future, with problems like drug abuse, crime, unemployment, sexism, racism, and local pollution, they show a large degree of pessimism, but they also expect some improvements. But when they view the future of the globe, their images are more pessimistic. War, environmental devastation, overpopulation, and famine are their main global fears, and they expect continuation or worsening of the global problems in the future (Brunstad, 2002; Eckersley, 1987, 2002; Gidley & Inayatullah, 2002; Head, 1997; Hicks, 1996; Lloyd & Wallace, 2004; Rubin, 2000).

The ROSE data give us the opportunity to separate environmental issues from other global challenges. What does our data say about youth's view of the future in relation

Table 1. Items addressing views of the future

D02. Environmental problems make the future of the world look bleak and hopeless
D07. We can still find solutions to our environmental problems
D14. I am optimistic about the future

to the environmental challenges? Are they optimistic or pessimistic? For example: one question in the instrument reads:

> To what extent do you agree with the following statements about problems with the environment (pollution of air and water, overuse of resources, global changes of the climate, etc.)?

The response scale goes from "Disagree" (coded 1) to "Agree" (coded 4). The value 2.5 constitutes the middle of the scale. This means that an average score of 2.5 may be considered as "neutral," meaning that the students on average neither agree nor disagree with the statement.

Three items intended to tap into the future images held by the respondents (Table 1), but they show weak inter-item correlations. Good measures in the affective domain can typically have inter-item correlations in the range of .30–.40 (Gable & Wolf, 1993), while the largest correlation coefficient between these three items is .20. The internal consistency is consequently unsatisfactory, and we cannot defend merging the items into one composite variable. In spite of these reliability problems with these three items, we will in the following report single-item scores in order to try to understand and validate the items. Results will be given for the total sample, as there are no noteworthy differences in girls' and boys' responses.

The responses in the three items do not portray youth as holding apocalyptic expectations to the future (Figure 1); maybe in contrast to what one could expect from the perspectives above drawn from literature. Among the three items, item D02 is the one most directly addressing the *environmental problems* and the *future of the world*. A mean score close to 2.5 implies that in average, the students neither agree nor disagree with the statement that the future of the world looks bleak and hopeless due to the environmental problems.

Item D14 ("I am optimistic about the future") was meant to be an opposite (negative) statement about the issue in D02. But this item text neither mentions explicitly the future *of the world* nor the *environmental problems*. The rather weak relationship between D02 and D14 (correlation coefficient −.20) implies that the two items do not function as positive and negative statements about the same underlying factor. Although item D14 is located under the questionnaire heading "Me and the environmental challenges," the response suggests that the students may have interpreted this item outside the global environmental context. The mean score in item D14 is 3.17, which is a fairly optimistic expression. We find it likely that the D14 item for some students may have denoted their *personal* future, which in other youth studies is found to be more optimistic and hopeful than the future of the globe (described above). This may be one of the reasons for the weak relationship between the two items.

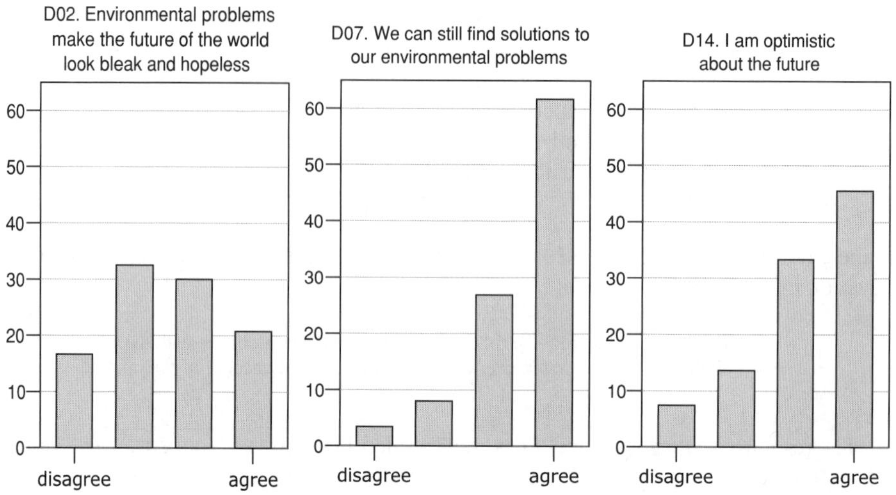

Figure 1. Views of the future (item D02, D07, and D14). Percentages of responses in the four response categories from "disagree" to "agree." (Due to some missing responses, the bars in each diagram do not add up to exactly 100%)

Item D07 displays a very hopeful profile (Figure 1) with mean value close to 3.5. The students strongly agree that we can still find solutions to our environmental problems. Question D02 and D07 are not correlated, so students agreeing that the future looks hopeless (item D02) may as well agree that we still can find solutions to the environmental problems (item D07). This should not be regarded as illogical or inconsistent responses, as a person may consider that even though it is *still not too late* to intervene and solve the problems (item D07), there is little hope that humanity actually will do so (item D02). This means that the two items may tap into substantively different issues (although they were designed for measuring the same).

In this section we wish to focus on environmental problems and students' images of the future of the globe. The above elaboration on the three questionnaire items addressing images of the future leads us to conclude that item D02 ("Environmental problems make the future of the world look bleak and hopeless") is probably the only item that was worded sufficiently precisely for tapping into this issue. In exploratory factor analysis we find this item loading on the same factor as "Nearly all human activity is damaging for the environment" (item D17) and "Science and technology are the cause of the environmental problems" (item G10). These items have in common a rather discouraging characteristic of the state of the world. Although item D02 does not show a pessimistic future, it is worth noticing and commenting on the 50% of the students, both girls and boys, agreeing (fully or partly) that the future of the world looks hopeless due to the environmental problems (Figure 1). It would be interesting to pursue these students through the data material, and characterise them in terms of other dimensions in the survey, but this would be beyond the scope of this chapter.

Are Youth Personally Engaged in the Environmental Protection Issue?

The sociologist Ulrich Beck sees risks as one of the main outcomes of globalization and technological and economical development and as contributing to the formation of a global "risk society." Until quite recently, people and societies were threatened by environmental risks that were unrelated to human activity, like drought, earthquakes, volcanoes, and storms. But the nature of the risks is changing. Globalisation presents us with new risks that are incalculable in origin and indeterminate in consequences. Today we meet ecological risks that are created by our own interventions with nature, social development, and development of S&T. The environmental problems are generated from human decisions and actions, but still they are diffuse in origin and have unknown and unfixed outcomes. Since the possible disasters are detached from individual responsibility, it is unclear who are responsible for finding answers to the problems and taking action to them (Beck, 1999).

Environmental protection is a prime concern in many facets of society. New taxes and duties are imposed, and people must deal with new laws and regulations, and adopt new practices. In this way public awareness about the environmental challenges is stimulated, but this does not necessarily develop positive and progressive attitudes. Our knowledge- and education society is frequently portrayed as a social system of technocracy, where citizens are alienated from responsibility and are lacking faith in their own opportunity to influence the societal development. They may believe that environmental problems should be left to the "experts."

Moreover, people in Western societies have access to large amounts of information, and today's young people have experienced the untrustworthiness of scientific discoveries and theories, and they know that knowledge may be contestable, short-lived, and after a period, outdated. Arguably, this makes people less convinced about "truths" and "facts," and creates a world in which individuals have become increasingly reflexive (Giddens, 1991). As we get new information and attain new knowledge, we consider and reconsider, design and redesign, and develop and redevelop ourselves, our beliefs, and our actions.

Descriptions as these are examples of some scientific, societal, and political complexities challenging environmental education for empowerment. As described in the rationale of the ROSE study (Schreiner & Sjøberg, 2004) some questionnaire items were initially developed with the intention of measuring various specified aspects of students' attitudes towards environmental protection. Again, in order to prevent response biases in the survey and thereby strengthen the reliability of the measures (Horan et al., 2003), we used the strategy of addressing the same issue with some items positively and some negatively worded. However, exploratory factor analysis did not confirm the existence of the intended underlying variables. The factor analysis sorted the items in two separate factors: positively worded statements in one factor and negatively worded items in another. This is a well-known effect of wording (see e.g. Horan et al., 2003; Marsh, 1996). However, the interpretation of this result is still undecided: Is the wording influencing a systematic irrelevant methodological artefact? Or are there substantive conceptual differences between negatively and positively worded items? Or may it be that this division of negatively and positively worded statements stems from response styles and personality traits of the respondents?

Table 2. Negatively worded items, describing a lack of concern for the environmental issue

D01. Threats to the environment are not my business
D03. Environmental problems are exaggerated
D08. People worry too much about environmental problems
D09. Environmental problems can be solved without big changes in our way of living
D13. Environmental problems should be left to the experts

We will not go any further into this discussion, as we regard it as closer to the field of psychometrics than science education. Through our conceptual analysis of the two item clusters, we found that the appearing factors were substantively meaningful. Consequently we will pursue the item clusters suggested by the factor analysis.

The negatively worded items seem to have in common a *lack of concern* for the environmental issue (Table 2). These items indicate that the environmental problems are exaggerated, that people cook up the problems, and that the individual puts a distance between herself or himself and the problems. If at all necessary, it is the task of somebody else to solve them. On the other hand, the positively worded items (Table 3) are describing a personal *involvement* in the issue (Item D07 from above is reappearing in this item cluster). These items describe attitudes towards the environmental problems suggesting that it is still possible to overcome the problems, a belief that every individual can make an important difference, and a willingness to act.

As the two groups of items show sufficient internal consistency, with Cronbach's alpha equal .69 for "Lack of concern" and .71 for "Involvement" respectively, we have computed two composite variables from the average scores of the items in each factor.

The average scores in these two composite variables for girls and boys expose students holding "socially accepted" attitudes to the environmental issue (Figure 2). The mean value for the variable "Lack of concern" is 2.1 for girls and 2.3 for boys. Mean values for both sexes are below the neutral 2.5 implying a lack of concern for the environment. Girls disagree somewhat stronger than boys. But on a disagree-scale ranging down to 1, these mean values show that neither of the sexes *strongly* disagree. That the students are lukewarm about the issue can be seen as confirmed by the scores on the "Involvement" variable: Mean scores close to 3.0 for boys and 3.1 for girls

Table 3. Positively worded items, describing a personal involvement in the environmental issue

D05. I am willing to have environmental problems solved even if this means sacrificing many goods
D06. I can personally influence what happens with the environment
D07. We can still find solutions to our environmental problems
D10. People should care more about protection of the environment
D12. I think each of us can make a significant contribution to environmental protection

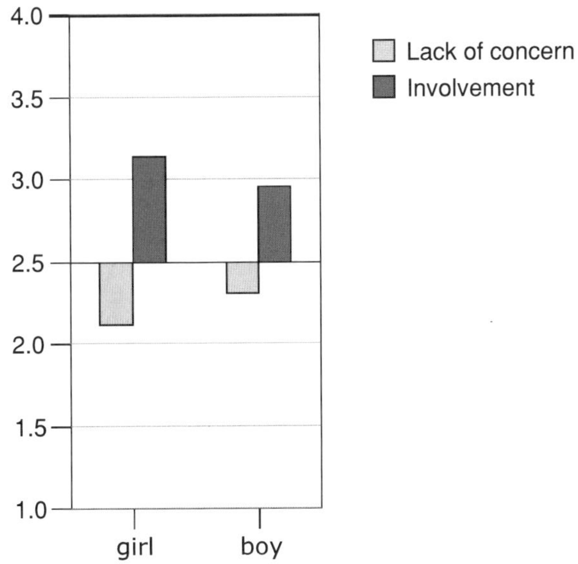

Figure 2. Means for girls and boys: Agreement (bars turning upwards) and disagreement (bars turning downwards) to the two composite variables "Lack of concern" (average of items in) and "Involvement" (average of items in)

(Figure 2) show little conviction to statements conveying personal involvement in the environmental issue.

The two composite variables are negatively correlated with a coefficient value of −.45. This relationship indicates that the two variables may be tapping into a common underlying factor on the topic of environmental protection. This brings us back into the methodological discussion mentioned above on positively and negatively worded statements. It may be that the substantive meaning of the one composite variable is close to the meaning of the other, although oppositely stated. We do, however, not see this discussion as important for our results. One conclusion we might draw from these two groups of items is that the students, to some extent, express concern about environmental issues; girls somewhat more than boys, but it does not seem to be a matter of great significance to them.

In a comparative Norwegian youth study, Brunstad (2002) found that although his informants had the relevant knowledge and insight, they had no feeling of having a possibility of influencing global development. Despite the moderate scores in our survey, our data tend not to support his finding.

Do Youth Find Environmental Protection Important for Society?

Thomas Ziehe sees late modernity as imposing a disruption between the individuals and the past and between the individual and her/his own culture (Ziehe & Stubenrauch, 1993). Late modern youth are culturally and socially liberated and freed from traditions and norms, and regardless of social background they have access to social goods such as

Table 4. Items for various goals for a society. The three first rows are items addressing the environmental issue (item L02, L04, and L09; Italics)

Items
L02. Protect untouched Norwegian nature
L04. Protect the environment against pollution
L09. Provide protection of our big predators
L03. More emphasis on medical research (e.g. on cancer and HIV/AIDS)
L16. Provide a society free from drugs
L11. Eradicate all forms of poverty and distress in Norway
L12. Lower taxes and duties
L14. Enhanced emphasis on education and better schools
L07. Enhanced emphasis on research on new technology
L01. Achieve high economic growth
L10. Prepare Norway for welcoming more refugees and immigrants
L15. Give economic support to poor countries
L13. Use gender quotas to have more women in senior appointments

education. But youth do not only have the *freedom* to choose between infinite options—they are *forced* to choose. And if they fail, they are responsible for the wrong decisions and choices. They try to find the area in which they have their interests and abilities, and to create their own good and meaningful life. Furlong and Cartmel (1997) see this struggle for finding one's own way ahead so demanding that modern youth show traits of self-centredness.

Turning again to our study: One part in the ROSE instrument contains an inventory of goals for a society, of which three items are addressing the environmental issue. The students were requested to indicate how important they found each of the goals for our society by ticking the appropriate box in a 10-point Likert scale. A response in the first category (Not Important) was coded 1, in the second 1.33, in the third 1.66, etc. From this follows that a response in the 10th category (Very Important) was coded 4.

In a universe of thinkable tasks and ambitions for a society, the limited numbers of items in this question (Table 4) are of course only a small and unrepresentative sample of possible goals. Nevertheless, we will in the following have a closer look at the scores of the three items regarding environmental protection by seeing them against a background of some other societal goals.

The three items related to the environmental issue (item L02, L04, and L09) show some internal consistency (Cronbach's alpha comes to .74). On a scale from 1 to 4, the mean score of the composite variable calculated from these three items lies close to 2.9, which is fairly high in a priority scale ranging from 1 to 4. This means that the students give high scores to the importance of environmental protection. Girls find this goal somewhat more important than boys (3.0 and 2.8 respectively), but the difference is small (although statistically significant for the composite variable). Furthermore, we find that, except from L10, nearly all items in this question achieve high scores (Figure 3). This means that the students responded that through the list of items in this question, most of them refer to issues that they find important for our society. Looking at the three "green" items in relation to the others, the overall picture is that environmental protection is not prioritised above other issues. Lower taxes, better schools, and a

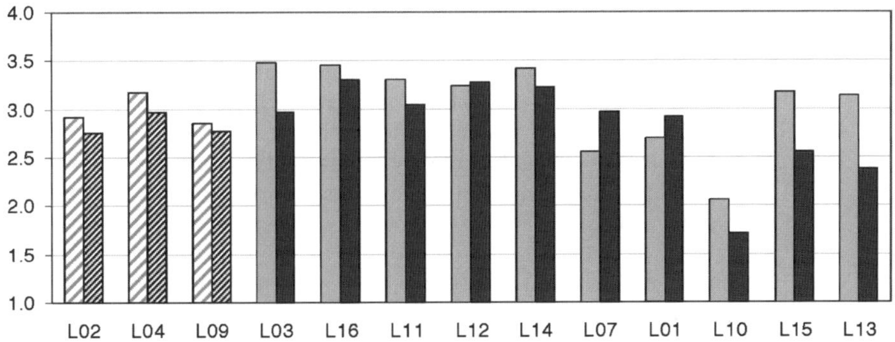

Figure 3. Important for our society: Mean scores for girls (light bars) and boys (dark bars) for the items in Table 4. Gender differences are not statistically significant for item L09, L12, and L16. The three leftmost bar clusters show scores for items addressing the environmental issue (item L02, L04, and L09; diagonal stripe pattern)

drug-free society (item L12, L14, and L16) receive higher scores from both girls and boys, while "gender quotas . . . " (item L13) achieve higher scores from the girls. These issues are closer and perhaps more relevant to the lives of Norwegian youth.

Girls give higher priority to health research (item L03) than to environmental protection. The health focus among the girls is in accordance with other questions in the ROSE questionnaire as well as with several youth studies (e.g. Bø, 1999), disclosing a considerable concern for health- and body-issues among young people in general and girls in particular.

Economic growth and research on new technology (item L01 and L07) are the only items where the boys give statistically significant higher scores than the girls. Other studies (e.g. Hellevik, 1996) find that boys are more influenced by a "race of affluence" than girls. In our survey, girls give less priority to national economic growth (item L01) than to environmental protection. They do, however, give high priority to *personal* economic prosperity (item L12).

A similar national–personal pattern can be read out of the items regarding foreign aid. The scores on the item for giving economic support to poor countries (item L15) are much higher than for welcoming more refugees and immigrants to Norway (item L10). The latter is more likely to threaten youth's personal success and happiness in life. Immigrants will be candidates in the same labour, love, and housing market. The cause of the low scores in item L10 is probably very complex, but one part of it may be that they perceive immigration as a threat for the success of their personal life.

How Interested are Norwegian Youth in Learning about Environmental Challenges?

The last facet of our concept for empowerment is the students' interest in learning about the environmental issue. For this purpose we will use a part of the questionnaire consisting of an inventory of possible topics to learn about, in total more than a hundred items. The students were requested to indicate on a 4-point scale how interested they were in learning about the various topics.

Table 5. Items with environmentally oriented topics to learn about

E05. What can be done to ensure clean air and safe drinking water
E06. How technology helps us to handle waste, garbage, and sewage
E03. The ozone layer and how it may be affected by humans
E04. The greenhouse effect and how it may be changed by humans
E33. Benefits and possible hazards of modern methods of farming
E19. Organic and ecological farming without use of pesticides and artificial fertilizers
E20. How energy can be saved or used in a more effective way
E21. New sources of energy from the sun, wind, tides, waves, etc.

Eight of the items are addressing the issue of environmental protection (Table 5). The group of items has a Cronbach's alpha of .88. Merged to one composite variable, this gives a mean value of 2.2. Both in an absolute and relative sense, this is a low score, which means that neither girls nor boys regard environmental protection as a matter of particular interest. The gender difference between the means of the composite variable is not statistically significant.

Seen in relation to the other topics in this part of the questionnaire, the environmental protection issue achieves low interest scores. For example, we find that topics referring to human body and health achieve much higher scores among the girls. We find statistically significant differences between girls and boys for the three items with an element of technology, invention, and/or energy (item for E06, E20, and E21, Figure 4). This resembles the gender difference pattern through the remaining of the hundred items; boys displaying more interests in technology and physics than girls. (This pattern is consistent with the picture from Figure 3.)

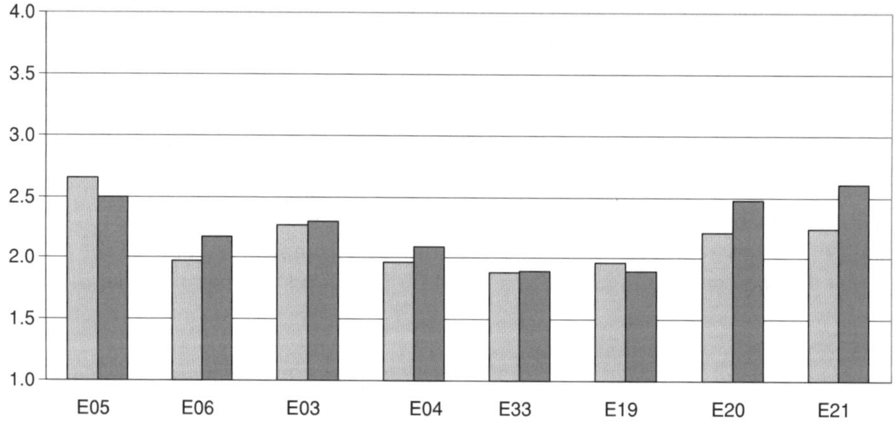

Figure 4. Mean scores for girls (light bars) and boys (dark bars) for interests in learning about the environmental protection topics in Table 5. Gender differences are statistically significant only for E06, E20, and E21.

In the section above we suggest that issues of *individual* relevance are of greater concern than issues associated with society and the globe. Mean scores for the eight items (in Figure 4) support this observation. The items closest connected to the life and the health of the individual (clean air and safe drinking water; item E05) achieves a relatively high interest score. This is surprising, given that a Norwegian identity is closely associated with the idea of "untouched nature," including clean air and drinking water. Again this may be understood as an apprehension of individual risk.

CONCLUSIONS AND IMPLICATIONS

We have argued that empowering students to act environmentally responsibly should be an important goal of education. By dividing the concept of "empowerment" into different aspects and subheadings, we have presented findings from the Norwegian data from the ROSE project.

Results from our analysis give us the overall impression that the students are only moderately engaged in the environmental issue, but the findings do not draw a very problematic portrait of youth. They do not seem unconcerned about (or alienated from) environmental problems, thinking that we still can find solutions to existing problems. Youth see individual contribution as important and think that they can personally influence what happens to the environment. But this engagement is not striking. Especially as the boys are only just within the bounds of the "politically correct" and the "socially accepted" attitudes. When it comes to the students' interest in learning about the environmental issue, they show little curiosity. Although many scientific subjects attract different interest among boys and girls, both girls and boys are equally lukewarm about learning about the topic. The result that may be perceived as the most disturbing is the fact that 50% of the students agree that environmental problems make the future of the world look bleak and hopeless.

Heilbroner (1995) found that from roughly 300 years ago until the second half of the 20th century, people in the West thought the future would be superior to the present. But as he came closer to our current period, he found that in advanced industrial and capitalist societies the visions of the future have been noticeably altered toward darker and more pessimistic images. According to Heilbroner, young people are growing up with a background of general public global future pessimism, and perhaps not surprisingly, they inherit these pessimistic views of the future and a limited sense of influencing it. Hicks (1996) refers to sources of expertise and experiences on how to engage students in envisioning a preferred future, and argues that these so far have not been sufficiently utilised by environmental educators. Among such resources he refers to the work of Jungk and Müllert (1987), Ziegler (1989), and Boulding (1988). Other valuable recourses may be found in Stapp (1996), Gidley & Inayatullah (2002), and the *World Yearbook of Education* (Hicks & Slaughter, 1998). A common denominator in many such strategies seems to be stimulating students' awareness of the future. People seem to know what they fear and what they might fight *against*, but are unsure about what they want for the future; what they would fight *for*. Empowering young people for action towards a better future might involve visualising alternatives as well as the goals of future struggles.

We suggest that environmental education should engage students in articulating, discussing, and eventually acting. In addition to going through *what we know* about a subject, the process should also address questions like (adapted from Hicks & Holden, 1995)

- What do we think, feel, hope, and fear in relation to the environmental problem? What do others who are involved think, feel, and say?
- Why do we and others think, feel, and act (or not act) in the way we do? What and who has influenced us? What is the history of this situation?
- Who has the power in this situation and how do they use it?
- What would things look like in a more just and sustainable future? What values will we use to guide our choices?
- What are the possible courses of action? What are others already doing? Which course of action is most likely to work for our preferred future?
- How shall we implement our plan of action in school, at home, and in the community? How shall we work together? Whose help might we need?
- How can we evaluate the outcome and measure our success?

In Western societies, studies find that the predominant factor showing environmental protection is the perception of risks, and to a smaller extent the value of nature *per se* (e.g. Skjåk & Bøyum, 1993). Moreover, our data indicate that societal and environmental matters, such as environmental risks or challenges facing a society, achieve a greater concern when such matters are connected to the personal life of the student. Some researchers interpret such findings in terms of characteristics of the so-called "here-and-now" generation or the "me"-generation. Øia (1995) offers another perspective. While older generations might conceive the environmental problems as "new," today's youth grow up with environmental disaster as a part of their life. The environmental challenges might have now become an accepted state of the world, but this does not necessary imply that they are unconcerned and unengaged in these matters.

A concern for self-actualisation and shaping ones own identity and happiness may be seen as a product of the prevailing ideas and the spirit of our time. Results from the international ROSE material show that, on average, youth from all countries surveyed, wish to be engaged in something they find important and meaningful, but we know that cultures have different meanings of the concept "important and meaningful." Western societies, for example, stress the importance of *living our lives to the fullest*. Our children are often brought up in a culture that encourages them to concentrate on things they like, find interesting, and are good at.

In order to achieve a deeper understanding of how schooling might serve to consolidate or change young people's attitudes towards environmental challenges, we would like to see more comparative research in this area. Cross-cultural comparisons in the ROSE material may, among other studies, provide us with more knowledge about such issues. But whatever these findings may say, we suggest that changing the prevailing level of interest, hope, and concern among youth is not easy. In order to design a suitable science and general education aimed at future citizens, some rethinking clearly needs to be done (and is, indeed, occurring). It is our hope that these efforts will continue and eventually contribute to an education for empowerment—and to the responsible management of our environment.

REFERENCES

Ary, D., Jacobs, L. C., & Razavieh, A. (1996). *Introduction to research in education*. Fort Worth: Harcourt Brace.

Beck, U. (1999). *World risk society*. Cambridge: Polity Press.

Bell, W. (1997). *Foundation of futures studies. Human science for a new era*. London: Transaction.

Bø, I. (1999). *Hadde ungdom det bedre før? En sammenlikning av ungdoms opplevelse av psykososiale problemer i 1959 og 1994* [*Were youth doing better before? A comparison of youth's perception of psychosocial problems in 1959 and 1994*] (Vol. 29). Stavanger: Høgskolen i Stavanger, Informasjonsenheten [Stavanger University College].

Boulding, E. (1988). *Building a global civic culture: Education for an interdependent world*. New York: Teachers College Press.

Brunstad, P. O. (2002). Longing for belonging: Youth culture in Norway. In J. Gidley & S. Inayatullah (Eds.), *Youth futures. Comparative research and transformative visions*. London: Praeger.

Eckersley, R. (1987). *Australian attitudes to science and technology and the future*: The Australian Commission for the Future.

Eckersley, R. (2002). Future visions, social realities, and private lives: Young people and their personal well-being. In J. Gidley & S. Inayatullah (Eds.), *Youth futures. Comparative research and transformative visions*. London: Praeger.

Furlong, A., & Cartmel, F. (1997). *Young people and social change: Individualization and risk in late modernity*. Buckingham: Open University Press.

Gable, R. K., & Wolf, M. B. (1993). *Instrument development in the affective domain. Measuring attitudes and values in corporate and school settings*. Boston: Kluwer Academic.

Giddens, A. (1991). *Modernity and self-identity. Self and society in the late modern age*. Cambridge: Polity Press.

Gidley, J., & Inayatullah, S. (Eds.). (2002). *Youth futures. Comparative research and transformative visions*. London: Praeger.

Head, S. (1997). *Futures in the class room—Student's viewpoints*. Paper presented at the World Futures Studies Federation XVth World Conference, Brisbane.

Heilbroner, R. (1995). *Visions of the future. The distant past, yesterday, today, tomorrow*. New York, Oxford: The New York Public Library and Oxford University Press.

Hellevik, O. (1996). *Nordmenn og det gode liv* [*Norwegians and the good life*]. Oslo: Universitetsforlaget.

Hellevik, O. (2002). *Forskningsmetode i sosiologi og statsvitenskap* [*Research method in sociology and political science*]. Oslo: Universitetsforlaget [Scandinavian University Press].

Hicks, D. (1996). Envisioning the future: The challenge for environmental educators. *Environmental Education Research, 2*(1), 101–108.

Hicks, D., & Holden, C. (1995). *Visions of the future, why we need to teach for tomorrow*. Staffordshire: Trentham Books.

Hicks, D., & Slaughter, R. (Eds.). (1998). *Futures Education. World Yearbook of Education*. London: Kogan Page.

Horan, P. M., DiStefano, C., & Motl, R. W. (2003). Wording Effects in Self-Esteem Scales: Methodological artifact or response style? *Structural Equation Modelling, 10*(3), 435–455.

Jenkins, E. W. (1994). Public understanding of science and science education for action. *Journal of Curriculum Studies, 26*(6), 601–611.

Jungk, R., & Müllert, N. R. (1987). *Future workshops: How to create desirable futures*. London: Institute for Social Inventions.

Klafki, W. (2001). *Dannelsesteori og didaktik—nye studier* (*Original: Neue Studien zur Bildungstheorie und Didaktik*). Århus: Klim.

Lloyd, D., & Wallace, J. (2004). Imaging the future of science education: The case of making futures studies explicit in student learning. *Studies in Science Education, 40*, 139–177.

Marsh, H. W. (1996). Positive and negative global self-esteem: A substantively meaningful distinction or artifactors? *Journal of Personality and Social Psycology, 70*(4), 810–819.

Nunnally, J. C. (1978). *Psychometric theory*. New York: McGraw-Hill.

Øia, T. (1995). *Apolitisk ungdom? Sjølbergingsgenerasjonen og politiske verdier* [*Apolitical youth? The self-rescuing generation and political values*]. Oslo: Cappelen Akademisk Forlag a.s.

Palmer, J. A. (1998). *Environmental education in the 21st century. Theory, practice, progress and promise*. London: Routledge.

Rubin, A. (2000). *Growing up in social transition: In search of a late-modernity identity.* Doctoral thesis, University of Turku, Finland.

Schreiner, C. (2004). *Report on organizing the ROSE survey in Norway, Oslo*. Available from: http://www.ils.uio.no/forskning/rose/ [access 2005-08-03].

Schreiner, C., Henriksen, E. K., & Hansen, P. J. K. (2005). Climate education—Empowering today's youth to meet tomorrow's challenges. *Studies in Science Education, 41*, 3–50.

Schreiner, C., & Sjøberg, S. (2004). *Sowing the seeds of ROSE. Background, Rationale, Questionnaire Development and Data Collection for ROSE (The Relevance of Science Education)—a comparative study of students' views of science and science education* (4/2004). Oslo: Department of Teacher Education and School Development, University of Oslo.

Skjåk, K. K., & Bøyum, B. (1993). *Undersøking om verdier, natur og miljø 1993* [*Attitudes towards the environment 1993*] (NSD report no 100). Bergen: ISSP (The International Social Survey Programme) and NSD (Norsk sammfunnsvitenskapelig datatjeneste).

Stapp, W., Wals, A. E. J., & Stankorb, S. L. (1996). *Environmental education for empowerment: Action research & community problem solving*. Iowa: Kendall/Hunt.

Toffler, A. (Ed.). (1974). *Learning For tomorrow. The role of the future in education.* New York: Random House.

Ziegler, W. (1989). *Envisioning the future: A mindbook of exercises for futures-inventors*. Denver, CO: Futures Invention Associates.

Ziehe, T., & Stubenrauch, H. (1993). *Ny ungdom og usædvanlige læreprocesser: kulturel frisættelse og subjektivitet* (*Originally published as: Plädoyer für ungewöhnliches Lernen, Ideen zur Jugendsituation, 1982*). København: Politisk Revy.

CHAPTER 6

SHERRY A. SOUTHERLAND & GALE M. SINATRA

THE SHIFTING ROLES OF ACCEPTANCE AND DISPOSITIONS IN UNDERSTANDING BIOLOGICAL EVOLUTION

INTRODUCTION

Consider the list of topics that can create challenges for science teachers in secondary science classrooms: Big Bang Theory, Genesis of Life, Biological Evolution, particularly the evolution of *Homo sapiens*, and even, HIV as the cause of AIDS. Although each of these represents a well-established scientific theory, each can cause some controversy. What is the root of the controversy? In some cases, these scientific theories contradict aspects of broader frameworks students have for understanding their world. In other cases, the theories fail to support ideas that students find appealing.

What makes controversial topics so difficult to teach? The most obvious problem may be found in the high emotional investment students bring to these topics—emotions that make engagement with the material very difficult for the teacher to orchestrate. But if we place these difficulties aside for a moment, we find that it can be difficult to *teach* controversial topics, because it can be difficult for students to *learn* such material. Trying to understand this difficulty has been the focus of our recent research.

For the layperson, the issue at the center of this analysis is clear-cut; one cannot accept what one does not know. The argument is that if a learner (and the learner's family) does not accept the idea of the Big Bang due to their strong religious traditions, then certainly one cannot expect much success in helping that learner understand the nuances of this theory of the origin of the universe. Science teachers often offer the opposite explanation. They sometimes argue that if students do not *understand* a scientific construct, then they have no basis upon which to accept that construct. Clearly there are some contradictions in these explanations of learning about controversial issues.

AFFECT AND MODELS OF CONCEPTUAL CHANGE

For years these commonsense explanations of students' difficulty with controversial topics were all but completely ignored in the academic literature of science learning. Indeed, early science education research centered on the role of prior knowledge and reasoning ability (constructs considered at that time to be nonaffective) in students' science learning. Recent work both within science education and in other domains such as educational psychology have begun to recognize the influence of affective constructs on learning. The Conceptual Change Model (CCM; Posner et al., 1982)

Steve Alsop (ed.) Beyond Cartesian Dualism, 69–78.

depicted a largely rational model of science learning, one that allowed little room for affect or other, extrarational considerations in learning. This changed somewhat with the revisionist approach to conceptual change as presented by Strike and Posner (1992), which recognized the limitations of the original CCM and acknowledged a much larger role of affect in cognition. However, it provided little understanding of affect beyond the acknowledgement of its import.

More recent models of cognition, such as the Cognitive Reconstruction of Knowledge Model (CRKM; Dole & Sinatra, 1998) and the Cognitive–Affective Model of Conceptual Change (CAMCC; Gregoire, 2003), feature strong affective components, components that include motivation, efficacy beliefs, implication of self, and intentions. The first of these two models, the CRKM, acknowledges the role of affect, yet it does not specifically describe the processes by which affective factors influence the learning of science. The second of these two, the CAMCC, takes affective models of conceptual change one step further by describing the role of affect in directing (not merely influencing) cognition. This model becomes particularly important for this discussion, as the CAMCC highlights the role of acceptance in learning. According to the CAMCC acceptance can prohibit the possibility of true conceptual change.

ACCEPTANCE, BELIEF, AND THE LEARNING OF EVOLUTION

Our goal is to further our understanding of the role of affect in cognition—with a particular emphasis on describing the role of acceptance and understanding in the learning of biological evolution. First, it is important to explain that we employ a distinction between *belief* and *acceptance* that is commonly used in the evolution education community (National Academy of Sciences, 1998; Smith, 1994; Smith et al., 1995; Smith & Scharmann, 1999). *Acceptance* of a construct is understood to require a systematic evaluation of evidence, while *belief* in a construct is understood to be based on personal convictions and opinions or an extrarational conviction. The goal of most science educators is to develop a learner's understanding of a construct, with the idea that instruction might impact her acceptance of that construct. We argue that it would be inappropriate to target a learner's personal *beliefs* in a science classroom, as beliefs are by definition outside empirical bounds, and thus outside the boundaries of scientific discussion. (For a more complete discussion of this position, see Southerland et al., 2001).

To explore the role of affect in science learning, we will discuss our work on the learning of a potentially controversial construct, biological evolution. We begin this discussion by describing some of the studies that have informed our research. A common position in the philosophical work regarding the learning of evolution is that a failure to accept a scientific construct precludes developing an understanding of it (Smith, 1994). Some theorists argue that a learner's acceptance of an idea must be addressed before he or she can come to understand the topic. If a learner fails to accept evolutionary theory, this disbelief will prevent him/her from attending to instruction. Thus, the argument that acceptance is pivotal to developing a scientific understanding of a construct (Cobern, 1994; Jackson, 2000; Meadows et al., 2000; Scharmann, 1990; Smith, 1994).

In contrast to this position, an early empirical study into understanding and belief [authors' wording] by Lawson and Worsnop (1992) focused on a learner's prior knowledge of evolution as a useful predictor of belief change. This analysis emphasized the role of reasoning abilities and their influence on belief and understanding. Less-skilled reasoners were more likely to hold nonscientific beliefs at the outset, were less likely to be strongly committed to evolution, and were less likely to change their acceptance during instruction. Prior knowledge and reflective reasoning skills were found to be the only two variables that accounted for a significant portion of the knowledge gain—acceptance did not. Lawson and Worsnop (1992) concluded that reflective reasoning skills allow one to come to understand evolution, and the resulting knowledge determines what one believes or accepts. Thus, knowledge is understood as the key to acceptance.

In yet a third position in this argument, some theorists argue that there is little or no relationship between acceptance of evolution and the understanding of this construct. This position is supported by both quantitative studies (Bishop & Anderson, 1990; Demastes-Southerland et al., 1995; Lord & Morino, 1993) and a qualitative study (Demastes-Southerland et al., 1996). This latter effort describes how a student with creationist views constructed a scientifically sophisticated understanding of evolutionary theory during a year of coursework. A second student in the study who had a stated acceptance of evolution developed only a superficial understanding of the theory. In support of this position, Dole et al. (1991) found no relation between students' stated belief in creationism and their ability to comprehend a text on evolution. These findings suggest that knowledge and acceptance are not as closely related as others have described.

DISPOSITIONS AS AFFECTIVE INFLUENCES ON LEARNING

Our focus is on examining how affect influences a learner's acceptance and understanding of evolution and previous research points to particular constructs that may play an important role. Mindful of the notion that the general public seems to characterize evolution as an inherently uncertain construct and one which conflicts with the religious beliefs of many, it would seem that a degree of willingness to examine one's beliefs and entertain alternative points of view may prove to be key factors in developing an understanding and acceptance of this theory.

Research from educational psychology demonstrates the influence of general tendencies toward thinking and learning, referred to as *dispositions* on student learning. Stanovich (1999) defines dispositions as "relatively stable psychological mechanisms and strategies that tend to generate characteristic behavioral tendencies and tactics" (p. 157). We believe that dispositions, which are similar to epistemological beliefs, (see Kardash & Sinatra, 2003) clearly have a strong affective component. Dispositions reflect a learner's tendencies to use knowledge and beliefs to direct actions toward particular learning goals and thus they have an important role in shaping the problem-solving and reasoning skills necessary for learning about complex, controversial topics (Stanovich, 1999). The work of Stanovich and his colleagues demonstrates that the dispositions to think in an open-minded fashion and to weigh new evidence against a personal belief help explain significant differences in problem-solving performance (Sá et al., 1999; Stanovich, 1999; Stanovich & West, 1997, 1998).

Some of the dispositions described in the literature that may well play a role in shaping the learning of evolution include *Activity Open-Minded Thinking* (Stanovich & West, 1997), a construct that assesses the tendency to be open to ideas that may conflict with one's own. Another relevant disposition, *Need for Cognition*, described by Cacioppo and colleagues (1996) measures the tendency to engage in effortful thinking. Finally, *Belief Identification* is a disposition that measures the degree to which an individual values stasis in their personal beliefs and believes that changing beliefs is a sign of weakness or disloyalty (Sá et al., 1999).

THE INFLUENCE OF DISPOSITIONS ON LEARNING ABOUT BIOLOGICAL EVOLUTION

Motivated by this previous research we designed a study that allowed us to examine the influence of affect on the learning of evolution (see Sinatra et al., 2003). In order to understand students' approach to a controversial topic such as human evolution, we examined their understanding and acceptance of this and less controversial theories. Thus, in this study of largely liberal education college students, we measured students' (a) understanding of evolution (using a measure by Settlage & Jensen, 1996) and understanding of photosynthesis (using a measure developed by Haslam & Treagust, 1987), (b) acceptance of human evolution, the evolution of nonhuman animals, and photosynthesis (using readings about photosynthesis/respiration, animal evolution, and human evolution), and (c) cognitive dispositions and epistemological beliefs (including the *Epistemological Belief Survey* developed by Wood & Kardash, 2002, and a 66-item questionnaire measuring various dispositions. See Appendix for sample questions). It is important to note that the students involved in this study were not biology majors, and most had little prior experience in biology courses.

We hypothesized that the relation between students' understanding and acceptance of a theory would depend on the theory in question. Therefore, we posited that there would be a positive relation between students' understanding of evolution and their acceptance of this theory. In contrast, we hypothesized that there would be no such relationship for learners' understanding of photosynthesis and their acceptance of this theory. Further, we hypothesized that affective constructs such as epistemological beliefs and learning dispositions would also influence acceptance and understanding of particular topics. Specifically, we hypothesized that students who viewed knowledge as changing and who have a disposition toward open-minded thinking would be more likely to accept the scientific explanation of human evolution, as well as be more likely to understand the construct. In contrast, we hypothesized that learning dispositions would not be related to acceptance or understanding of the noncontroversial topics.

Contrary to our expectations, in our statistical analysis of the data from 93 students, we found the only significant relation between knowledge and reported acceptance was for photosynthesis. There was no relation between knowledge and acceptance for animal or human evolution. More in accordance with our expectations, cognitive dispositions and epistemological beliefs were shown to be predictive of acceptance, but only for the controversial theory of human evolution.

There are several points of interest in our findings. First, our data showed no evidence of a relationship between understanding evolution and its acceptance. This stands in sharp contrast to the view that it is the content of one's knowledge that serves as a barrier to acceptance (Lawson & Worsnop, 1992). However, our findings do correspond to others in the field (Bishop & Anderson, 1990; Demastes-Southerland et al., 1995, 1996; Lord & Marino, 1993). Indeed, for this group of students it seems that students may have an understanding of evolutionary theory without accepting its validity, or alternatively, they may accept the validity of the construct based upon a very poor understanding of it.

Cognitive dispositions, however, were related to both knowledge and to acceptance of human evolution. The disposition scales measure a student's recognition of the need to change one's thinking, with a value placed on knowledge change. As an example, the dispositional measure, *Actively Open-minded Thinking* reflects a willingness to be open to new ideas, an affective construct that may be necessary for successful knowledge change. Our findings indicate that this measure was predictive of both knowledge of and acceptance of human evolution. However, dispositions were not related to knowledge of or acceptance of the less controversial theories of animal evolution or photosynthesis.

These findings support the notion that a willingness to intentionally entertain knowledge change (a construct with a strong affective component) affects both knowledge of and acceptance of scientific theories, when the learning of those theories evokes strong emotions. Our data suggest that learning about human evolution clearly does evoke strong emotions for some students.

Our data demonstrate that the relationship between understanding, acceptance, and learning dispositions varies with the content in question. Dispositions were related to a learner's understanding of evolution, but not their understanding of photosynthesis. Knowledge of photosynthesis was related to acceptance of that theory, but no relation was found between knowledge of evolution and acceptance of evolution. Thus, we see that the nature of the intersection between acceptance, understanding, and learning dispositions varies with the domain in question. We would argue that the degree of controversy of the domain plays an important role in shaping this interaction, precisely because students' emotions are involved when learning about such topics.

THE MEDIATING ROLE OF KNOWLEDGE

After completing the study described above, we were puzzled regarding one central finding—the lack of relationship between students' understanding of evolution and their acceptance of it. Although other researchers had arrived at similar conclusions (Bishop & Anderson, 1990; Demastes-Southerland et al., 1995; Lord & Morino, 1993), we remained skeptical. Indeed each of these studies, including our own, had one common factor—data were gathered from students with very limited knowledge of evolution. In retrospect, it seems likely that the disjuncture between knowledge and belief we observed may have been due to participants' limited knowledge of evolution. Perhaps their knowledge simply was not strong enough to influence their acceptance. We designed a second study to test the notion that when knowledge is sufficient, knowledge and acceptance would be related. (For a complete description of this study, see Sinatra et al., 2004).

Study II employed the methods from the first study, but with this effort we compared two populations of college students, education majors (56 students with an average of one college-level biology course completed) and biology majors (56 students with an average of 6.8 college-level biology courses completed). Again, three sets of constructs (dispositions, acceptance, and understanding) were measured for three theories with differing degrees of controversy (photosynthesis, animal evolution, and human evolution).

To establish that the biology majors and education majors differed in their knowledge of photosynthesis and biological evolution, the groups were compared on their performance on the two knowledge measures (photosynthesis and evolution). Statistical analysis revealed that the two groups differed significantly in their knowledge of photosynthesis, with the biology majors scoring 63% correct and the education majors scoring 50% correct. There were even more dramatic differences between the two groups for the measure of biological evolution, with the biology majors scoring 70% correct and the education majors scoring 44% correct, again a statistically significant difference.

Did the groups differ significantly in ways other than knowledge? The education majors were primarily females (77%), whereas males were in the majority among the biology majors (61%), a statistically significant difference. Also, the education majors were slightly older as a group, again a statistically significant difference. Most important for the results of the present study, however, the two groups did not differ in their dispositions or epistemological beliefs. Indeed, the difference between the education majors and the biology majors on the dispositions scale was not significant. The groups' mean ratings on each of the five epistemological belief subscales are also strikingly similar, with no significant differences between the groups.

A series of regression analyses were used to examine the relation of knowledge, beliefs, and dispositions to students' acceptance of the three theories in question. The findings for photosynthesis revealed that dispositions were not predictive of students' reported acceptance of this scientific theory. This supports the findings of Study I. For animal evolution, the findings also echoed the patterns established in Study I. For both education and biology students, knowledge and dispositions were not significantly related to students' acceptance of animal evolution. This pattern held true even though the biology students' knowledge of and degree of acceptance of animal evolution was higher than the education majors. Knowledge of biological evolution still failed to predict acceptance of animal evolution for these students. Thus, even with two groups that differed significantly in terms of their knowledge of biological evolution, knowledge did not differentially predict reported acceptance of animal evolution.

In contrast to the results for these less controversial theories, the results for human evolution differed between the two groups of students. While education majors' dispositions predicted their acceptance of this controversial construct, for biology majors, it was *knowledge* that significantly predicted acceptance of human evolution. Thus, our results for the education majors mirrored those of Study I, understanding was not related to acceptance of evolution for the education major. However, for the biology majors who had a much higher level of knowledge, understanding was found to be a significant predictor of students' acceptance of human evolution.

In summary, it seems that the degree to which affective constructs are evoked when learning about scientific constructs depends on the degree of controversy of the topic in question, and the amount of knowledge of the topic students bring to the learning situation. For students not well acquainted with biology, acceptance is not associated with understanding for biological evolution, while for students well immersed in biology, there is a strong relationship between a learners' acceptance and their understanding.

Considering both studies leads to the suggestion that knowledge must reach a critical level to influence students' acceptance of ideas. When knowledge is limited, when the content is perceived as controversial and/or tentative, and when the ideas are related to firmly held beliefs, then affective constructs such as the willingness to change one's thinking may play a more significant role in determining acceptance than background knowledge.

MESHING OUR FINDINGS WITH CURRENT MODELS OF CONCEPTUAL CHANGE

Both the CRKM (Dole & Sinatra, 1998) and the CAMCC (Gregoire, 2003) acknowledge a role of dispositions and acceptance in shaping learners' understanding. The CAMCC describes how acceptance of a construct plays a very important role in shaping eventual understanding of potentially controversial content—a role that may be mitigated by learning dispositions. Our findings support the assertion that acceptance and dispositions may shape the way in which learners engage with material, and thus may serve to influence their eventual understanding of that material. However, in only a limited number of instances in our work was acceptance found to be related to understanding. That relationship was observed only for knowledgeable biology majors' acceptance of a potentially controversial topic. The difference between our findings and those portrayed in the CAMCC may likely be the degree of controversy of the content in question.

Our findings do suggest, however, that models such as the CRKM and the CAMCC, which portray conceptual change as a process driven by affective factors, not simply influenced by them, to be on the right track in their acknowledgement that affective factors play a *determining* role in whether or not change occurs. Both models, and the results of our research, suggest that whether a learner engages deeply or superficially with material will depend on the learner's willingness to do the cognitive work necessary for high engagement, as described in the CRKM. Affective factors, such as anxiety, perceived threat, resistance to change, self-efficacy, epistemological beliefs, as well as traditionally identified factors such as ability and prior knowledge all contribute to whether the learner will engage with the content in a manner that can promote change. Both the CRKM and CAMCC suggest if a learner is confronted with these affective factors directly, in the course of learning, than possibilities of using affective factors as a positive force for change are enhanced.

Our findings demonstrate that mapping the intersection of affective and cognitive factors is a very complex undertaking, and one that is sensitive to both the degree of controversy associated with the content of instruction and the sophistication of prior knowledge of the learner. Indeed, it is a dynamic process—a learner's knowledge influences how acceptance impacts understanding, but the nature of that influence shifts as the degree of background knowledge changes.

We believe continued focus on the intersection of affective and cognitive factors is called for, as we begin to recognize that learning is not solely determined by the characteristics of the content in question or unconscious attributes of the learner (i.e., reasoning ability, background knowledge). Instead, we have begun to recognize that some learning occurs because of a learner's conscious intention to learn—that affective constructs such as goals, beliefs, motivation, and openness to new ideas shape how a learner engages with material, and therefore shapes a student's eventual understanding of that material (Sinatra & Pintrich, 2003).

Despite the difficulties of modeling both the affective and cognitive components of learning science content, a continued focus on affect as central to science learning will create a better appreciation of the complexities of learning difficult and sometimes controversial content. Understanding the influence of affective factors on learning more completely is necessary if we are to design instruction to support students' intentional learning of complex and controversial science content.

APPENDIX: SAMPLE ITEMS FROM DISPOSITION MEASURES

In order to measure participants' learning dispositions, a 66-item, 5-point Likert scale questionnaire was used. The dispositional measures chosen were based on the work of Stanovich and his colleagues (Sá et al., 1999; Stanovich, 1999) and developed by a variety of researchers. Measures included *Actively Open-minded Thinking (AOT) and Belief Identification* (Sá et al., 1999; Stanovich & West, 1997), *Dogmatism* (Troldahl, & Powell, 1965), *Categorical Thinking* (Epstein & Meier, 1989), *Absolutism* (Erwin, 1983), *Values* (Costa & McCrae, 1992), and *Need for Cognition* (Cacioppo et al., 1996). Two sample items for each scale appear below. For complete scales, see the original citations.

Actively Open-minded Thinking Scale Sample Items

Changing your mind is a sign of weakness.

If I think longer about a problem I will be more likely to solve it.

Dogmatism Scale Sample Items

Of all the different philosophies that exist in the world there is probably only one which is correct.

No one can talk me out of something I know is right.

Absolutism Scale Sample Items

A professor's job is to communicate the facts to his or her students.

Right and wrong never change.

Values Scale Sample Items

I believe letting students hear controversial speakers can only confuse and mislead them.

I believe that laws and social policies should change to reflect the needs of a changing world.

Categorical Thinking Scale Sample Items

There are basically two kinds of people in this world, good and bad.

I think there are many wrong ways, but only one right way, to almost anything.

Belief Identification Scale Sample Items

One should disregard evidence that conflicts with your established beliefs.
Certain beliefs are just too important to abandon no matter how good a case can be made against them.

Need for Cognition Scale Sample Items

I would prefer complex to simple problems.
I usually end up deliberating about issues even when they do not affect me personally.

REFERENCES

Bishop, B. A., & Anderson, C. W. (1990). Student conceptions of natural selection and its role in evolution. *Journal of Research in Science Teaching, 27,* 415–427.

Cacioppo, J. T., Petty, R. E., Feinstein, J., & Jarvis, W. (1996). Dispositional differences in cognitive motivation: The life and times of individuals varying in need for cognition. *Psychological Bulletin, 119,* 197–253.

Cobern, W. W. (1994). Belief, understanding, and the teaching of evolution. *Journal for Research in Science Teaching, 31*(5), 583–590.

Costa, P.T., & McCrae, R. R. (1992). Normal Personality Assessment in Clinical Practice: The NEO Personality Inventory. *Psychological Assessment, 4*(1), 5–13.

Demastes-Southerland, S., Good, R., & Peebles, P. (1996). Students' conceptual ecologies and the process of conceptual change in evolution. *Science Education, 79*(6), 637–666.

Demastes-Southerland, S., Settlage, J., & Good, R. (1995). Students' conceptions of natural selection and its role in evolution: Cases of replication and comparison. *Journal of Research in Science Teaching, 32*(5), 535–550.

Dole, J. A., & Sinatra, G. M. (1998). Reconceptualizing change in the cognitive construction of knowledge. *Educational Psychologist, 33*(2/3), 109–128.

Dole, J., Sinatra, G. M., & Reynolds, R. (1991, December). *The effects of strong beliefs on text processing: The case of evolution and creationism.* Paper presented at the National Reading Conference Annual Meeting, Palm Springs, CA.

Epstein, S., & Meier, P. (1989). Constructive thinking: A broad coping variable with specific components. *Journal of Personality and Social Psychology, 57,* 332–350.

Erwin, T. D. (1983). The Scale of Intellectual Development: Measuring Perry's scheme. *Journal of College Student Personnel, 24,* 6–12.

Gregoire, M. (2003). Is it a challenge or a threat? A dual-process model of teachers' cognition and appraisal processes during conceptual change. *Educational Psychology Review, 15*(2), 147–179.

Haslam, F., & Treagust, D. F. (1987). Diagnosing secondary students' misconceptions of photosynthesis and respiration in plants using a two-tier multiple choice instrument. *Journal of Biological Education, 21*(3), 203–211.

Jackson, D. F. (2000). *Shifting the relationship between personal and professional beliefs and practices with regard to evolution and religion: Three years of feedback from prospective middle school science teachers.* A paper presented at the annual meeting of the National Association for Research in Science Teaching, New Orleans, LA.

Kardash, C. A., & Sinatra, G. M. (2003). *Epistemological beliefs and dispositions: Are we measuring the same construct?* Paper presented at the annual meeting of the American Educational Research Association, Chicago, IL.

Lawson, A. E., & Worsnop, W. A. (1992). Learning about evolution and rejecting a belief in special creation: Effects of reflective reasoning skill, prior knowledge, prior belief, and religious commitment. *Journal of Research in Science Teaching, 29*(2), 143–166.

Lord, T., & Marino, S. (1993). How university students view the theory of evolution. *Journal of College Science Teaching, XXII*(6), 353–357.

Meadows, L., Doster, E., & Jackson, D. F. (2000). Managing the conflict between evolution and religion. *The American Biology Teacher, 62*(2), 102–107.

National Academy of Sciences. (1998). *Teaching about evolution and the nature of science.* Washington, DC: National Academy Press.

Posner, G., Strike, K., Hewson, P., & Gertzog, W. (1982). Accommodation of a scientific conception: Toward a theory of conceptual change. *Science Education, 66*(2), 211–227.

Sá, W., West, R. F., & Stanovich, K. E. (1999). The domain specificity and generality of belief bias in reasoning and judgement. *Journal of Educational Psychology, 91*(3), 497–510.

Scharmann, L. C. (1990). Enhancing an understanding of the premises of evolutionary theory: The influence of a diversified instructional strategy. *School Science and Mathematics, 90*(2), 91–100.

Settlage, J., & Jensen, M. (1996). Investigating the inconsistencies in college student responses to natural selection test questions. *Electronic Journal of Science Education.* Available at: <http://unr.edu/homepage/jcannon/ejse/settlage.html> (Version current at January 11, 2005).

Sinatra, G. M., & Pintrich, P. R. (2003). *Intentional conceptual change*. Mahwah, NJ: Lawrence Erlbaum.

Sinatra, G. M., Southerland, S. A., & Demastes, J. (2004). *A little knowledge is a dangerous thing: Using beliefs and dispositions to make judgments about the validity of scientific theories.* Paper presented at the annual meeting of the National Association for Research in Science Teaching, Vancouver, BC.

Sinatra, G. M., Southerland, S. A., McConaughy, F., & Demastes, J. (2003). Affective and intentional influences on understanding: Intersections of understanding, acceptance/belief, and epistemology for biological evolution. *Journal of Research in Science Teaching, 40*(5), 510–528.

Smith, M. U. (1994). Counterpoint: Belief, understanding, and the teaching of evolution. *Journal for Research in Science Teaching, 31*(5), 591–597.

Smith, M. U., & Scharmann, L. C. (1999). Defining versus describing the nature of science: A pragmatic analysis for classroom teachers and science educators. *Journal for Research in Science Teaching, 83,* 493–509.

Smith, M. U., Siegel, H., & McInerney, J. D. (1995). Foundational issues in evolution education. *Science & Education, 4,* 23–46.

Southerland, S. A., Sinatra, G. M., & Matthews, M. (2001). Belief, knowledge, and science education. *Educational Psychology Review, 13*(4), 325–351.

Stanovich, K. E. (1999). *Who is rational? Studies of individual differences in reasoning.* Mahwah, NJ: Lawrence Erlbaum.

Stanovich, K. E., & West, R. F. (1997). Reasoning independently of prior belief and individual differences in actively open-minded thinking. *Journal of Educational Psychology, 89,* 342–357.

Stanovich, K. E., & West, R. F. (1998). Individual differences in rational thought. *Journal of Experimental Psychology: General, 127,* 161–188.

Strike, K. A., & Posner, G. J. (1992). A revisionist theory of conceptual change. In R. A. Duschl and R. J. Hamilton (Eds.), *Philosophy of science, cognitive psychology, and educational theory and practice* (pp. 147–176). New York: State University of New York.

Troldahl, V., & Powell, F. (1965). A short-form dogmatism scale for use in field studies. *Social Forces, 44,* 211–215.

Wood, P., & Kardash, C.A. (2002). Critical elements in the design and analysis of studies of epistemology (pp. 231–260). In B. K. Hofer & P. R. Pintrich (Eds.), *Personal epistemology*. Mahwah, NJ: Lawrence Erlbaum.

SECTION TWO: TEACHING, LEARNING, AND AFFECT

OVERVIEW

In Section two, our discussions shift from learners' emotions to the phenomena of teaching and learning. The following four chapters explore the influence of affect in both formal and informal educational settings.

An individual perspective is again evident as emotions are linked to cognition through the psychological construct of motivation, more specifically motivational beliefs, such as self-efficacy, interest, and task value. In these analyses, considerable weight is placed on the personal/environmental interaction—the interface between individuals' intrinsic motivational preferences and motivation generated extrinsically by particular instructional techniques and/or environmental features. The overriding desire is mainly to enhance cognitive performance (demonstrably increased quality and/or quantity of learning) and the focus is a transitional phase in which emotion is used to understand, and evaluate the effects of instructional techniques.

The authors mine the literature to analyse how particular instructional practices increase motivation and as a consequence enhance student achievement. Some approach this from the perspective of motivational theory (Chapter 7), while others, focus more closely on a particular pedagogical approach (e.g. practical work, see Chapter 8) or more broadly free-choice learning environments (see Chapter 9). In the latter two cases, theory and empirical work are used to better understand practice, while in the former case, a synthesis of existing literature forms of the basis of a series of generalisations about practice (i.e. what approaches previous empirical work suggests might serve to increase situational motivation). Although cognitive gains are mostly the ultimate goal, in the context of free-choice learning (Chapter 9) these discussions are more progressive advocating the importance of affective goals—the validity of gains in individual motivation as the possible outcome of a visit to an informal education setting, a museum or a travelling exhibit, for instance.

More specifically

Chapter 7 provides an overview of the seminal work of the science education group at the University of Michigan. In their analysis, Christine Rhee Bonney and colleagues subdivide motivational beliefs into those that *mediate* instructional strategies and achievement and those that *moderate* instructional effects. A critical review of the literature brings to the fore the influence that such beliefs as self-efficacy, task value, interest, and achievement goals have on cognitive outcomes. Future work, they suggest, might profitably adopt a mediational/moderator framework to specifically investigate efficacious instructional practices. Interested readers can find further details of the Michigan Math and Science Partnership—Motivation Assessment Program (MSP – MAP) on the Web, at http://www.mspmap.org.

Steve Alsop (ed.) Beyond Cartesian Dualism, 81–82.

In some jurisdictions, practical work has been a dominant part of science education for over a century and yet as Jerry Wellington laments it has largely escaped evaluative scrutiny. Does it work? Is it worth the investment? Can it be used more effectively? Only recently have such questions been raised, let alone explored. Wellington's consummate discussion derives from an analysis of literature and previously unpublished data extracted from a UK national survey. Although clearly much more research is needed, the evidence presented suggests that practical work has an important general role in motivation. Wellington advocates the use of an ecological framework to holistically investigate the *mediating* and *moderating* role of practical work (using the language of the preceding chapter). Such analysis, he contends, should associate judgements of efficacy with socioaffective, as well as cognitive gains.

Chapter 9 steps through the "window" and into the looking glass of free choice learning. Informal learning contexts, the author Lynn Dierking suggests, might offer a unique environment in which to investigate the intricate interplay of affect and cognition. Dierking's essay contemplates enhanced engagement as a product of autonomy and multisensory experience. Evidence of empirical richness resides in two cases: an HIV/AIDS travelling exhibition and a Family learning initiative at the Children's Museum, Indianapolis. Here, the dual obligation of free-choice learning encounters are apparent—the need to conceptualise learners as emotional, social, and cultural beings, as well as cognitive beings. Future agendas, Dierking suggests, might look across an evolving narrative of experience as "people engage in, support and facilitate learning across the different aspects of their lives."

The final chapter in this section offers a quite different focus and theoretical perspective. In Chapter 10, Zembylas draws on the work of social constructivist and poststructural thinking to explore teachers' emotions. For social constructivists and poststructuralists, emotions are viewed as cultural constructions that are situated and serve particular social ends. Rather than seeing emotions as situated with the minds of individuals, theorists examine the ways in which emotions are embodied within particular cultures and social practices, such as language, institutional structures, and social relationships. More specifically, Chapter 10 has two clearly articulated aims: (i) To evaluate current research on the role of emotions in teaching in general, and (ii) propose some possible directions for future investigations of teacher emotion in science education. In an edited collection that largely assumes an individual perspective, it is important to underscore the significance of this chapter both in terms of its focus and in terms of its theoretical perspective.

CHAPTER 7

CHRISTINA RHEE BONNEY, TONI M. KEMPLER, AKANE ZUSHO, BRIAN P. COPPOLA, & PAUL R. PINTRICH

STUDENT LEARNING IN SCIENCE CLASSROOMS: WHAT ROLE DOES MOTIVATION PLAY?

INTRODUCTION

One of the major goals of science education today is the attainment of scientific literacy, which includes deeper conceptual understanding of key scientific principles and ideas, the ability to apply scientific knowledge in real-life contexts, as well as the ability to identify problems and conduct scientific inquiry (American Association for the Advancement of Science, 1994; Marx et al, 1997). In examining the antecedents of scientific literacy, one fruitful avenue of research has been the work on student cognition. In particular, this work has underscored the affordances and constraints of prior knowledge influencing conceptual change. These cognitive models of learning have focused mainly on factors such as encoding, automatization, and metacognitive strategies, which have been found in laboratory studies to play a critical role in the conceptual change process. However, there is a need to also consider noncognitive factors such as students' motivational beliefs, especially when examining students' cognitive engagement in academic classrooms (Pintrich et al., 1993; Zusho et al, 2003). Accordingly, the purpose of this chapter is to discuss the value of motivation within science education.

It is important to note that we conceptualize motivation in this chapter more as a process, rather than as a product. Drawing on recent research from social–cognitive and situated perspectives, we stress the multidimensional nature of motivation and examine how motivational processes are influenced by classroom contextual factors. In short, we do not consider motivation to be a general trait, with some students more and others less motivated along a general quantitative continuum. Rather, we assume student motivation to be situated and changeable as a function of instruction, tasks, and activities that take place in a classroom.

In considering the relation between motivation and achievement, we propose two general ways in which motivational beliefs can influence positive academic outcomes (Linnenbrink & Pintrich, 2002). First, motivational beliefs can be thought to "mediate" the relation between certain instructional strategies and achievement (see Figure 1a). For example, the implementation of a new inquiry-based curriculum can result in students becoming more interested in science, which ultimately could lead to higher levels of achievement. In short, this view assumes that "good" instruction should lead to more adaptive motivational processes, which should in turn lead to positive academic outcomes (Stipek, 2001).

Steve Alsop (ed.) Beyond Cartesian Dualism, 83–97.

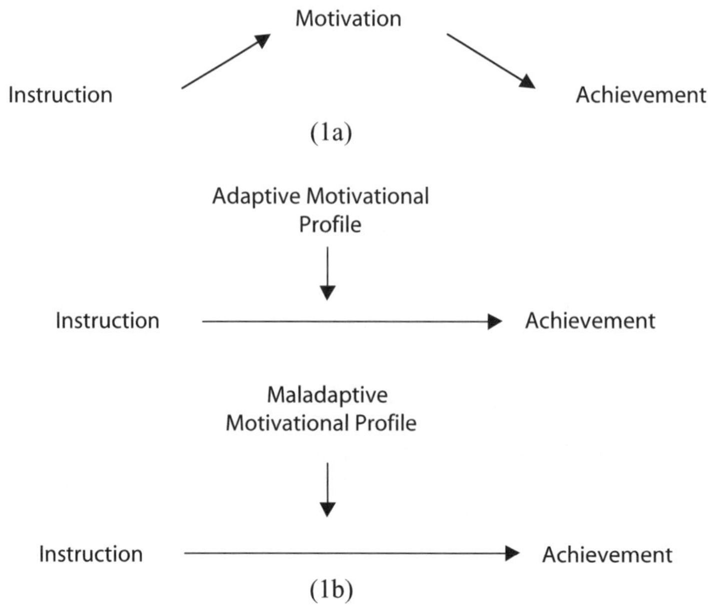

Figure 1. (a) How motivational beliefs mediate the relation between instruction and achievement; (b) How motivational beliefs moderate the relation between instruction and achievement

In contrast to this mediational model, motivational beliefs can also be thought of as "moderators" of instructional effects (see Figure 1b). For example, one could conceive that challenging, more constructivist-based methods of science instruction could be most beneficial to those students who typically like and value science (i.e., those with adaptive motivational profiles), in comparison to those students who are less strategic or who fail to recognize the value of scientific pursuits (i.e., those with maladaptive motivational profiles).

In discussing these two alternative accounts, we focus our analysis on the motivational beliefs of self-efficacy, task value, interest, and achievement goals. Self-efficacy refers to students' beliefs that they have the resources and confidence to do the tasks in the classroom (Bandura, 1997). Self-efficacy is *not* "self-esteem," which refers to an undifferentiated affective evaluation of the self. Rather, self-efficacy concerns specific social–cognitive judgments of one's capabilities.

Task value beliefs are another important motivational component (Eccles et al., 1998). In contrast to the general question of "Am I capable of performing this task?" which is part of self-efficacy beliefs, task value beliefs focus on the general question of "Why do I want to do this task?" In short, task value beliefs concern beliefs about the importance and utility of the subject matter domain, in this case science.

Personal interest refers to an individual's attraction, general liking, and enjoyment of a specific activity or domain (Pintrich & Schunk, 2002). While Eccles and her colleagues (1998) typically consider personal interest (or what they refer to as intrinsic value) under the umbrella of task value beliefs, we believe personal interest to be an important outcome in its own right, especially in light of the marked efforts across various reform efforts to increase students' levels of interest in the sciences (Koller et al., 2001). For this reason, we have chosen to discuss this motivational belief separately in this chapter.

Finally, we consider students' achievement goals as possible mediators or moderators of achievement. Specifically, we examine the role of mastery goals, or goals focused on learning and understanding, as compared to performance goals, or goals focused on outperforming others.

Starting with the next section, we discuss each of these motivational beliefs in detail, beginning with the mediational argument, followed by a consideration of how these beliefs moderate the relation between instruction and achievement. We conclude each section with an examination of how factors such as age, gender, and ethnicity influence these motivational beliefs.

SELF-EFFICACY

As mentioned previously, the construct of self-efficacy includes students' judgments about their capabilities to accomplish certain goals or tasks by their actions in specific situations, as well as their beliefs about their agency in the course (Bandura, 1982, 1986; Pajares, 1997; Schunk, 1991). This approach implies a relatively situational or domain-specific construct rather than a global personality trait. In an achievement context, self-efficacy beliefs are primarily influenced by the demands of a task, including skills that are perceived to be necessary for working on the task, students' general perceptions about a subject domain, as well as prior performance on similar tasks. In turn, self-efficacy beliefs have been found to influence activity choice, as well as students' overall levels of effort and persistence (Bandura, 1997). Generally, researchers have shown that it is more adaptive to have higher efficacy beliefs. For example, students who believe that they are capable of adequately completing a task and have more confidence in their ability to do so, typically display the highest levels of academic achievement and also engage in academic behaviors that promote learning (Bandura, 1997; Schunk, 1991; Zusho et al., 2003).

Self-Efficacy as Mediator

In terms of a mediational argument, there are a number of ways in which instructional strategies can influence students' self-efficacy beliefs. First, the task itself can have a profound effect on students' perceptions of their capabilities to complete a task. For example, research suggests that students may make lower than usual self-efficacy judgments in the face of challenging activities (Pintrich & Schunk, 2002). As tasks require students to evoke higher order strategies and integrate a variety of new and learned skills, students may perceive these tasks as extending beyond their present capabilities, therefore engendering feelings of anxiety or perceptions that they are ill-prepared to engage in these tasks.

However, there are ways to offset such potentially detrimental beliefs. For example, without necessarily simplifying the cognitive demands of a task, instructors can help maintain self-efficacy levels by breaking down the activity into more manageable components, helping students set appropriate goals, and encouraging them to think about effective ways to approach the task (Turner et al., 2002). Teachers can also convey the importance to students that science is indeed learnable and that it is possible to increase one's knowledge and skill of science by employing specific strategies (Zusho et al., 2003). For instance, teachers can first demonstrate or model the use of a new strategy, and then continue to monitor and support students' attempts to synthesize concepts by posing questions that gradually increase in level of difficulty, by encouraging students to draw connections among ideas, by clarifying any concepts that remain unclear, and by offering feedback about students' current levels of understanding.

Self-Efficacy as Moderator

As new directions in the field of science aim to provide students with opportunities to engage in investigations and solve real-world problems, students frequently confront a higher degree of challenge. Specifically, these reform-minded curricular programs often require students to apply a range of strategies and engage in more difficult tasks such as long-term projects and scientific inquiry (Krajcik et al., 1998). In addition, students are also expected to flexibly employ both domain-general and science-specific skills while developing their conceptual understanding of the science content.

A critical question is posed concerning whether the science instruction advocated by science reforms has a moderating effect on the relation between self-efficacy and student achievement. It is reasonable to expect, for example, that these more challenging instructional practices that engage students in complex investigations may prove most beneficial for those students with higher self-efficacy beliefs. In fact, a moderating relation suggests that it is the interaction between this high self-efficacy and inquiry-based science instruction that enables students to access the most advanced levels of conceptual understanding. Highly efficacious students stand to benefit from these science reforms because they are able to adapt more readily to the challenges introduced through these changes in instruction. In contrast, for those students who are low in self-efficacy, these same science curricula may have no effect or a detrimental effect on student achievement outcomes, such that these students might avoid the task and decrease their subsequent levels of engagement. This question raises important implications for science educators and leaders of these science reform efforts concerning the challenges raised by introducing inquiry methods of teaching, and underscores the importance of maintaining adaptive motivational beliefs.

Individual Differences in Self-Efficacy

Perhaps the most striking findings concerning individual differences in self-efficacy beliefs relate to the role of gender. Specifically, gender differences have been noted in students' ratings of their self-efficacy beliefs, as well as their general perceptions of academic competence, with males usually displaying higher levels than females, especially in traditionally male-dominated fields such as science and mathematics (Eccles et al., 1998). Moreover, it has been found that this disparity in the competency ratings of boys and girls becomes even more pronounced following puberty. Researchers have typically accounted for such findings in terms of gender-role stereotyping and gender socialization. Eccles and her colleagues (1998) have asserted that while boys may hold higher competency ratings for athletics and mathematics, females often display higher ratings for subjects such as reading, English, and social studies. The magnitude of these differences, however, varies depending on the extent to which boys and girls actually endorse cultural values regarding gender-related superiority in these domains.

Interestingly, despite such gender disparities, investigators have found very little evidence suggesting that males actually outperform females academically (Eisenberg et al., 1996). This raises the issue of calibration, that is, the extent to which students' ratings of their motivational beliefs such as self-efficacy, accurately reflect their true level of motivation and achievement as measured by some external, objective standard. Some researchers have argued that males typically overestimate how well they think they will perform on future tasks, while females generally underestimate their abilities (Eccles et al., 1998). Likewise, a similar phenomenon has been noted among African American students. Despite generally low levels of achievement, African American students have been found to report remarkably high expectations for success, thus leading theorists to conclude that like males, African American students may tend to inflate their ratings of their academic abilities (Graham, 1994).

Although limited, some work has also been done to examine how students' reports of their self-efficacy vary by age. For example, Zusho and Pintrich (2003) have proposed that the development of self-efficacy could exhibit contrasting patterns. On the one hand, one could expect task specific judgments of efficacy to increase over time as students gain more skill and expertise at a specific task (e.g., Shell et al., 1989). In contrast, as academic tasks get increasingly difficult, one could also expect a drop in efficacy ratings. The important point is that the development of specific self-efficacy judgments should be tied to actual experience on the task, not necessarily to general developmental trends or age-related changes, in contrast to research on more global competence perceptions, which shows a very reliable and steady decline over time (Eccles et al., 1998).

TASK VALUE

In an effort to explain why students want to succeed on a task, three components of task value have been proposed as influencing the achievement behavior within a particular content domain; these include attainment value, intrinsic value, and utility value (Eccles, 1983; Eccles & Wigfield, 1995). Attainment value refers to the importance or salience that students place on the task. Intrinsic value (i.e., personal interest) relates to general enjoyment of the task or subject matter, which remains more stable over time. Finally, utility value concerns students' perceptions of the usefulness of the task, in terms of their daily life or for future career-related or life goals. As previously mentioned, we believe that interest, which closely parallels Eccles' conceptualization of intrinsic value (Eccles, 1983), is an important motivational outcome in its own right, and will therefore be discussed in greater detail in the next section.

Task Value as Mediator

The relation between instructional practices and achievement may be mediated through students' task value, such that students enrolled in science classrooms that emphasize the utility and importance of science activities will increasingly perceive the content and activities to be useful and will subsequently become more involved, use more effective learning strategies, and perform better than students in traditional science classrooms. In line with certain science reform initiatives, teachers may implement inquiry-based instruction, in a way that enhances the perceived utility for all participating students. For example, these science reform efforts attempt to increase task value by incorporating opportunities for students to engage in authentic scientific activities, by relating science content to students' everyday experiences, and by making connections to community-related issues (Marx et al., 1997). In this manner, students gain access to the science content by investigating questions and issues that they perceive to be useful and valuable to their lives outside of school.

Students' task value can influence achievement-related outcomes within the domain of science through the extent to which they choose to engage during science instruction. For example, those students who hold high utility and attainment value beliefs for the domain of science may opt to continue their coursework in the sciences, and increase their level and quality of involvement on specific tasks. In contrast, those students who have conceptions of science as unrelated to the real world, for example, may not try to master the content to the same degree. Although there is little evidence that increased task value leads directly to higher achievement, the two are positively correlated, suggesting that it is still an important construct that teachers should consider and encourage among their students (Pintrich & Schunk, 2002).

Task Value as Moderator

Task value can also serve as a possible moderator of the relation between instruction and achievement. Specifically, students who have had prior experience with such inquiry-based instruction as described above may enter the classroom with higher levels of task value compared with students who have been in more traditional science classrooms. Such students may be more receptive to instruction that highlights the utility and importance of scientific tasks. Conversely, students who may not see how science is useful or important may be especially vulnerable to declines in interest, engagement, and ultimately, achievement.

Individual Differences in Task Value

Similar to self-efficacy, gender differences have also been noted in value components of motivation. For example, Eccles, Wigfield, and their colleagues have uncovered several gender-related differences in adolescents' valuing of certain subjects (Eccles et al., 1993; Wigfield & Eccles, 1992). Specifically, they found that girls typically valued instrumental music and English more than boys, while boys valued sports more than girls. Surprisingly, male and female adolescents did not differ in the relative value they attached to mathematics until they reached high school, when boys reported valuing math more than girls. Given these domain differences, gender can moderate the general negative trajectory in the development of value or interest, but the effect will vary by domain with males more "at risk" in writing or English, and females more at risk in mathematics and sports over time.

From a developmental perspective, there is evidence that children's reports of their value for academic domains decline across the school years. It appears that elementary students report valuing schoolwork more than middle-school and high-school students in general (Eccles et al., 1998). Despite this overall general decline, however, the developmental progression is believed to be analogous to that of self-efficacy and competence beliefs. That is, from a task and individual differences perspective, there is reason to believe that this general trend may vary depending on the task and situation. For example, there are clearly tasks and activities that students grow to value over time, especially when they begin to have some choices about how to spend their time (Pintrich & Zusho, 2001). Thus, while there may be an overall general decline, there can be activity- or task-specific increases in task value over the course of development.

INTEREST

As mentioned earlier, students' interest refers to an attraction, enjoyment of, or general liking for a particular domain or academic subject (Pintrich & Schunk, 2002). Researchers often distinguish between personal and situational interest. Personal interest can be conceptualized as a personality trait or a relatively stable, enduring disposition of the person that is normally directed toward a specific activity or task, such as sports, music, or reading (Pintrich & Schunk, 2002). While personal interest is more stable and internal to the learner, another conceptualization of interest involves characteristics of the context that make a particular task or activity interesting (interestingness), resulting in situational interest, or the psychological state of being interested in a specific activity (Pintrich & Schunk, 2002). The study of interestingness has generally focused on text, and features of text passages that might be found interesting to groups of people (Hidi & Anderson, 1992; Krapp et al., 1992); however, this could also be applied to other domains, such as science activities. In this sense, the interestingness of a particular task or content should result in situational interest that is not unique to one individual, but can be influential across many individuals. Clearly this has implications for teachers who, by increasing the interestingness of a task or lesson, can foster situational interest in their students.

Interest as Mediator

Generally, interest is assumed to be positively related to learning and, therefore, achievement because of the heightened concentration, use of learning strategies, and challenge-setting that often results from interest (Hidi et al., 1992; Krapp & Fink, 1992; Renninger & Hidi, 2000). Therefore, situational interest can be studied as a potential mediator of achievement. For example, Harackiewicz et al. (1997) investigated the relation between college psychology students' motivation, interest, and achievement, and found a positive relation between interest and reports of higher levels of self-efficacy, which were consequently related to higher grades in the class. Similarly,

Laukenmann et al. (2003) found that when physics instructors organized instruction into an initial acquisition phase and a second practice phase, during which instruction was relaxed and focused on students experiencing physics competency in small successes, situational interest was significantly related to performance on physics assessments.

Also within the domain of science, Black and Deci (2000) demonstrated how, in a college-level organic chemistry course, students' perceptions of how much their instructor supported their autonomy positively predicted students' level of interest and enjoyment in the course. Students in these classes were encouraged to be active learners by attending formal lectures, as well as intensive study groups led by advanced students, and were provided mentoring and social support (see Black & Deci, 2000, for review). This student-centered approach to teaching organic chemistry demonstrates how teachers' instructional methods can influence students' situational interest in science classes, which then impacts their level of achievement.

Interest as Moderator

Personal interest, as an internal characteristic of the student, may not be as influenced by instruction as situational interest; however, it may serve as a moderator of the relation between good instruction and achievement such that students with higher personal interest may perform better in classes than students with lower personal interest in the material. There is a noteworthy relation between interest and knowledge, which may influence how students process and learn information. For example, students may have an easier time reading and processing a text passage that is interesting compared to a noninteresting passage, because they are more likely to remember details of the interesting passage (Renninger, 1992). However, the more students become "experts" in a particular domain, the role interest plays in learning would be expected to change, such that the student should be less likely to be distracted by "seductive details" and more likely to identify essential points.

It is clear that knowledge and interest do not necessarily go hand-in-hand (Renninger & Hidi, 2000). For example, Hidi and Anderson (1992) reviewed a study in which students reading topics that were either unfamiliar or very familiar to them reported low levels of interest, suggesting that an optimal level of interest may result from moderate levels of prior knowledge on a topic. Interestingly, however, students who read about high-interest low-knowledge topics (Space Travel) performed worse on a writing task than students who read about low-interest high-knowledge topics (Living in a City). This suggests that even with low levels of interest, a students' broad knowledge base could buffer the detrimental effects of a boring task. Therefore, although it may be beneficial from a motivational standpoint to engage children in interesting activities, it is imperative to keep in mind that they must first be armed with sufficient knowledge to carry out the task effectively. Without proper instruction and an adequate knowledge base from which to complete academic tasks, motivational factors such as interestingness and "seductive details" may actually be detrimental to students' learning.

Finally, Schiefele et al. (1992) conducted a meta-analysis of research investigating interest as a predictor of achievement. They reviewed 121 independent random samples from 18 different countries, with grade levels ranging from 5th to 12th grade and covering nine different subject matters. Across the entire sample of studies, interest accounted for about 10% of the observed achievement variance. Therefore, although many studies have found evidence suggesting a positive relation between interest and achievement, it is important to consider other factors such as prior achievement that explain some of the other 90% of achievement variance not accounted for by interest.

Individual Differences in Interest

Surprisingly little research has been conducted on developmental and gender differences in interest. According to Renninger (1992), children as young as 3 years old exhibit relatively stable personal interests, operationalized as playing with particular toys. Young children's interests tend to influence more activities and tasks than for older children, possibly due to the previously discussed relation between interest and prior knowledge. Older children and adults have more experiences with a wider array of tasks and may already have ideas of what is interesting to them and what is not. The decline in interest has also been attributed to the fact that older children, as a function of general development, begin to explore new and different interests, which distract them from interests in school-related activities (Renninger & Hidi, 2000). Interestingly, Schiefele et al.'s meta-analysis (Schiefele et al., 1992) revealed that younger students' interest—achievement correlations were comparable to those of older students.

Explorations of gender differences in interest development have yielded some interesting results. Harackiewicz et al. (1997) found that among college students in an introductory psychology class, females reported enjoying the class more than males. Among elementary-school students, Renninger (1992) found a gender-by-interest interaction on performance of math problems, such that boys made more mistakes on problems rated as uninteresting, whereas girls made more mistakes on interesting problems. She explains this result by suggesting that interest may facilitate boys' learning by keeping them motivated and engaged, but that interest may serve as a distraction for girls, interfering with their learning and problem solving.

Similarly, Schiefele et al. (1992) found in their meta-analysis that female students' academic performance was less related to their reports of interest than was male students' performance. Most research on gender differences in interest has focused on the domain of the natural sciences. Consistently, males are more interested (and perform better) in natural sciences while females report higher interest and earn higher grades in literature and English. These differences also tend to increase as students get older. Schiefele et al. (1992) discuss possible explanations for such differences in interest and performance by domains, such as sex differences in basic abilities that would influence learning in these particular domains; however, no conclusive results have been obtained.

There has also been a relative dearth of empirical research on ethnic differences in interest development, and there has yet been little theoretical justification for such an examination; however, this is not to say that research on other individual differences are unjustified. Future research needs to investigate the relation of interest and achievement in different academic domains, as well as examine possible gender and ethnic differences.

ACHIEVEMENT GOALS

Another important component of student motivation concerns their general achievement goals, or their goals for academic learning and in classroom contexts. The general distinction between mastery and performance goals contrasts students who are mastery oriented and focused on learning and understanding (similar in some ways to intrinsic motivation) and those students who are performance oriented and focused on doing better than others in terms of grades or other outcomes that invite interpersonal comparisons (Pintrich, 2000a, 2000b).

There is a need to examine the role of these goals within the context of various classroom environments. Many mathematics and science classrooms, especially those at the secondary and postsecondary levels, use competitive grading systems and it may be adaptive in these contexts for students to adopt performance goals. As schools and teachers make changes to improve instruction, these grading systems may change, and the goals may function differently.

Achievement Goals as Mediators

In investigating the link between teachers' instructional practices and students' motivation, much of the research has focused on the type of environment or classroom context teachers establish for their students. The extent to which students perceive that the class structure makes a specific goal salient may influence their own goal orientation toward the course material (Ames, 1992; Ames & Archer, 1988). Research has suggested that by manipulating characteristics of the classroom environment (e.g., tasks, evaluation and assessment procedures, authority), teachers can make it more likely that their students will perceive either a mastery or a performance goal focus (Ames, 1992; Pintrich & Schunk, 2002).

The second half of the mediational model involves the relation between the goals students adopt and achievement. When examining direct relations, researchers often find that mastery goals are unrelated to students' performance (cf. Church et al., 2001); however, there are indirect relations, in which mastery goals are associated with students' use of deeper learning strategies, which are then, in turn, related to higher achievement (Pintrich, 2000a). With respect to performance goals, existing research on the relation between this goal and achievement has been somewhat mixed, with some researchers finding that performance goals are positively related to classroom performance (Pintrich, 2000a; Skaalvik, 1997; Wolters et al., 1996), while others have found no significant relation (Zusho et al., 2003).

Achievement Goals as Moderators

Classroom goal structures may influence the types of personal goals students adopt; however, these structures may also interact with students' personal goals to influence achievement. One hypothesis about how students' personal goals may moderate the relation between teachers' classroom goal structures and student achievement involves the idea of student–environment match. Under this hypothesis, performance-oriented students who are in a classroom performance context, and mastery-oriented students who are in a classroom mastery context, would be expected to exhibit adaptive academic outcomes, presumably due to the congruence between their personal goals and the perceived goal structure in the classroom (Linnenbrink, 2004).

Recent research has begun to suggest that students are not necessarily limited to adopting either mastery or performance goals, but can adopt multiple goals (Pintrich, 2000a). Traditionally, personal mastery goals have been seen as adaptive and personal performance goals have been seen as maladaptive; however, given recent research that indicates the relation between performance goals and higher performance, this multiple goals perspective includes the idea that students may be most advantaged if they endorse high levels of both goals (Barron & Harackiewicz, 2001; Pintrich, 2000a). In line with this perspective and the student—environment match hypothesis, it has been suggested that students who endorse high levels of both mastery and performance goals may have the best academic prospects because they can selectively choose the most appropriate goal to pursue at a given time, based on the perceived demands of the classroom (Barron & Harackiewicz, 2001).

Another hypothesis of how personal goals can moderate the effect of classroom goal structure on achievement involves the notion of a "buffering" effect. As mentioned earlier, some science classrooms, especially at the secondary and postsecondary levels, employ the use of grading curves, which may serve to increase the perceived level of competition among students, perhaps making more salient to students a classroom performance goal structure. If a mastery-oriented student is placed in a performance-oriented classroom, the student's personal goals might interact with the instructional context by "buffering" the student from any negative effects of a performance context.

Individual Differences in Achievement Goals

Limited work has been done investigating the extent to which individual differences between students influence the adoption, and influence, of achievement goals. Developmentally, younger students might be more likely to adopt mastery goals because they hold incremental theories of intelligence, believing that intelligence can change over time with effort; conversely, as students get older, it is suggested that they develop more entity theories of intelligence, believing their intelligence and ability level will remain stable over time regardless of effort (Dweck & Elliott, 1983). While this belief may not necessarily be conducive to a performance goal adoption, students are less likely to believe that deeper understanding of the material will improve their ability.

The empirical support that gender influences achievement goal adoption is mixed. It has been suggested that girls underestimate their academic ability more than males, having lower expectations for success and a greater tendency to attribute failure to a lack of ability and success to luck or other uncontrollable factors (Eccles et al., 1998; Eisenberg et al., 1996; Pintrich & Schunk, 2002). This might imply that girls would be more likely than boys to adopt performance goals and less likely to adopt mastery goals; however, this is not necessarily the case. Gender stereotypes of boys being more aggressive and competitive than girls would suggest that boys are more likely to be performance oriented than girls. Also, stereotypes of girls being more nurturing and caring than boys might translate into girls not wanting to hurt someone's feelings by beating them, thus being less likely to adopt a performance goal than boys (Eccles et al., 1998; Eisenberg et al., 1996).

The research examining cultural variation in goal adoption has been limited, although Maehr and Nicholls (1980) have proffered evidence suggesting that the goals driving achievement-related behavior might differ based on how cultures view success and failure. There has also been some speculation related to possible differences between Asian- and Anglo-American students' adoption of specific learning goals (Zusho & Pintrich, 2003). For example, citing home and schooling practices focused on the exhortation of effort and perseverance, some researchers have argued that Asians may be more mastery oriented than Westerners (Holloway, 1988; Whang & Hancock, 1994). The empirical support for such a claim, however, has been scant. Whang and Hancock (1994), for example, found no significant differences between Asians and non-Asians in their mean level ratings of both mastery and performance goals, although they did find mastery and performance goals to be significant predictors of mathematics performance. Lee et al. (2003), on the other hand, observed that American students place greater emphasis on mastery goals and that Asian students endorse higher levels of performance goals.

CONCLUSIONS AND FUTURE DIRECTIONS

Clearly, motivation can play a significant role in student learning in science classrooms. While it is a complex phenomenon with numerous variables influencing student learning, in this chapter, we focused on four motivational beliefs, namely self-efficacy, task value, interest, and achievement goals. There are two general ways that these beliefs can influence achievement: as mediators between instructional activities and achievement, and as moderators of instructional effects (Linnenbrink & Pintrich, 2002; Pintrich, 1999; Pintrich et al., 1993). Under the mediational pathway, students' beliefs are influenced by various instructional activities, and these beliefs in turn lead to increased student achievement. In contrast, under the moderator pathway, the beliefs or strategies interact with instruction to influence student achievement.

Future directions in science education research should focus on both the mediational and moderator models, specifically on how teachers can utilize knowledge of these pathways and relations to understand how their instructional practices may affect student achievement. At the University of Michigan, we are beginning to undertake a large-scale, multiyear project, funded by the National Science Foundation, called the Math and Science Partnership—Motivation Assessment Program (MSP–MAP). Our goals for this project include the following: (1) to develop and make available reliable, valid, and practical tools to assess student motivational beliefs for mathematics and science, as well as learning strategies and epistemological beliefs, that can be used by mathematics and science classroom teachers to evaluate the effectiveness of their interventions; (2) to increase teachers' knowledge about the role of these beliefs and strategies as either mediators or moderators of instruction and how they are related to student achievement in mathematics and science, in a manner that informs the design and evaluation of interventions; and (3) to assist teachers by providing information about how student beliefs and strategies, and their linkages to student achievement, generalize or may differ as a function of gender, age, ethnicity, and socioeconomic status. We hope that by further investigating the impact of student motivation on achievement, math and science teachers may adopt and incorporate various instructional practices in their classrooms, which could lead to optimal levels of motivation among students, and subsequent higher achievement.

REFERENCES

American Association for the Advancement of Science. (1994). *Benchmarks for science literacy*. New York: Oxford University Press.

Ames, C. (1992). Classrooms: Goals, structures, and student motivation. *Journal of Educational Psychology, 84*, 261–271.

Ames, C., & Archer, J. (1988). Achievement goals in the classroom: Students' learning strategies and motivation processes. *Journal of Educational Psychology, 80*, 260–267.

Bandura, A. (1982). Self-efficacy mechanisms in human agency. *American Psychologist, 37*, 122–147.

Bandura, A. (1986). *Social foundations of thought and action: A social cognitive theory.* Englewood Cliffs, NJ: Prentice-Hall.

Bandura, A. (1997). *Self-efficacy: The exercise of control*. New York: W.H. Freeman.

Barron, K. E., & Harackiewicz, J. M. (2001). Achievement goals and optimal motivation: Testing multiple goal models. *Journal of Personality and Social Psychology, 80*, 706–722.

Black, A. E., & Deci, E. L. (2000). The effects of instructors' autonomy support and students' autonomous motivation on learning organic chemistry: A self-determination theory perspective. *Science Education, 84*, 740–756.

Church, M. A., Elliot, A. J., & Gable, S. L. (2001). Perceptions of classroom environment, achievement goals, and achievement outcomes. *Journal of Educational Psychology, 93*, 43–54.

Dweck, C. S., & Elliott, E. S. (1983). Achievement motivation. In P. H. Mussen (Series Ed.) & E. M. Heatherington (Vol. Ed.), *Handbook of child psychology: Vol. 4. Socialization, personality, and social development* (4th ed., pp. 643–691). New York: Wiley.

Eccles, J. (1983). Expectancies, values, and academic behaviors. In J. T. Spence (Ed.), *Achievement and achievement motives* (pp. 75–146). San Francisco: W. H. Freeman.

Eccles, J. S., & Wigfield, A. (1995). In the mind of the actor: The structure of adolescents' achievement task values and expectancy-related beliefs. *Personality and Social Psychology Bulletin, 21*, 215–225.

Eccles, J., Wigfield, A., Harold, R. D., & Blumenfeld, P. (1993). Age and gender differences in children's self- and task perception during elementary school. *Child Development, 64*, 830–847.

Eccles, J. S., Wigfield, A., & Schiefele, U. (1998). Motivation to succeed. In N. Eisenberg (Ed.), *Handbook of child psychology: Vol. 3. Social, emotional, and personality development* (5th ed., pp. 1017–1095). New York: Wiley.

Eisenberg, N., Martin, C. L., & Fabes, R. A. (1996). Gender development and gender effects. In D. C. Berliner & R. C. Calfee (Eds.), *Handbook of educational psychology* (pp. 358–396). New York: Macmillan.

Graham, S. (1994). Motivation in African Americans. *Review of Educational Research, 64*, 55–117.

Harackiewicz, J. M., Barron, K. E., Carter, S. M., Lehto, A. T., & Elliot, A. J. (1997). Predictors and consequences of achievement goals in the college classroom: Maintaining interest and making the grade. *Journal of Personality and Social Psychology, 73*, 1284–1295.

Hidi, S., & Anderson, V. (1992). Situational interest and its impact on reading and expository writing. In K. A. Renninger, S. Hidi, & A. Krapp (Eds.), *The role of interest in learning and development* (pp. 215–238). Hillsdale, NJ: Erlbaum.

Hidi, S., Renninger, K. A., & Krapp, A. (1992). The present state of interest research. In K. A. Renninger, S. Hidi, & A. Krapp (Eds.), *The role of interest in learning and development* (pp. 433–446). Hillsdale, NJ: Erlbaum.

Holloway, S. D. (1988). Concepts of ability and effort in Japan and the United States. *Review of Educational Research, 58*, 327–345.

Koller, O., Baumert, J., & Schnabel, K. (2001). Does interest matter? The relationship between academic interest and achievement in mathematics. *Journal of Research in Mathematics Education, 32*, 448–470.

Krajcik, J., Blumenfeld, P. C., Marx, R. W., Bass, K. M., & Fredricks, J. (1998). Inquiry in project-based science classrooms: Initial attempts by middle school students. *The Journal of the Learning Sciences, 7*, 313–350.

Krapp, A., & Fink, B. (1992). The development and function of interests during the critical transition from home to preschool. In K. A. Renninger, S. Hidi, & A. Krapp (Eds.), *The role of interest in learning and development* (pp. 397–429). Hillsdale, NJ: Erlbaum.

Krapp, A., Hidi, S., & Renninger, K. A. (1992). Interest, learning, and development. In K. A. Renninger, S. Hidi, & A. Krapp (Eds.), *The role of interest in learning and development* (pp. 3–26). Hillsdale, NJ: Erlbaum.

Laukenmann, M., Bleicher, M., Fus, S., Gläser-Zikuda, M., Mayring, P., & von Rhöneck, C. (2003). An investigation of the influence of emotional factors on learning in physics instruction. *International Journal of Science Education, 25*, 489–507.

Lee, C., Tinsley, C., & Bobko, P. (2003). Cross-cultural variance in goal orientations and their effects. *Applied Psychology: An International Review, 52*, 272–297.

Linnenbrink, E. A. (2004). Person and context: Theoretical and practical concerns in achievement goal theory. In P. R. Pintrich & M. L. Maehr (Eds.), Motivating students, improving schools: The legacy of Carol Midgley (Vol. 13, pp. 159–184). Oxford: Elsevier.

Linnenbrink, E. A., & Pintrich, P. R. (2004). Multiple goals, multiple contexts: The dynamic interplay between personal goals and contextual goal stresses. In S. Volet & S. Jarvela (Eds.), *Motivation in learning contexts: Theoretical advances and methodological implications* (pp. 251–269). Amsterdam: Elsevier Science.

Linnenbrink, E. A., & Pintrich, P. R. (2002). The role of motivational beliefs in conceptual change. In M. Limon & L. Mason (Eds.), *Reconsidering conceptual change: Issues in theory and practice* (pp. 115–135). Dordecht, The Netherlands: Kluwer Academic.

Maehr, M. L., & Nicholls, J. G. (1980). Culture and achievement motivation: A second look. In N. Warren (Ed.), *Studies in cross-cultural* psychology (Vol. 3, pp. 221–267). New York: Academic Press.

Marx, R. W., Blumenfeld, P. C., Krajcik, J. S., & Soloway, E. (1997). Enacting project-based science. *The Elementary School Journal, 97*, 341–358.

Pajares, F. (1997). Current directions in self-efficacy research. In M. Maehr & P. R. Pintrich (Eds.), *Advances in motivation and achievement* (Vol. 10, pp. 1–49). Greenwich, CT: JAI Press.

Pintrich, P. R. (1999). Motivational beliefs as resources for and constraints on conceptual change. In W. Schnotz, S. Vosniadou, & M. Carretero (Eds.), *New perspectives on conceptual change* (pp. 33–50). Oxford, United Kingdom: Elsevier Science.

Pintrich, P. R. (2000a). Multiple goals, multiple pathways: The role of goal orientation in learning and achievement. *Journal of Educational Psychology, 92*, 544–555.

Pintrich, P. R. (2000b). The role of goal orientation in self-regulated learning. In M. Boekaerts, P. R. Pintrich, & M. Zeidner, (Eds.), *Handbook of self-regulation* (pp. 451–502). San Diego, CA: Academic Press.

Pintrich, P. R., Marx, R. W., & Boyle, R. A. (1993). Beyond cold conceptual change: The role of motivational beliefs and classroom contextual factors in the process of conceptual change. *Review of Educational Research, 63*, 167–199.

Pintrich, P. R., & Schunk, D. H. (2002). *Motivation in education: Theory, research, and applications* (2nd ed.). Upper Saddle River, NJ: Merrill Prentice-Hall.

Pintrich, P. R., & Zusho, A. (2001). The development of academic self-regulation: The role of cognitive and motivational factors. In A. Wigfield and J. Eccles (Eds.), *The development of achievement motivation* (pp. 250–284). San Diego, CA: Academic Press.

Renninger, K. A. (1992). Individual interest and development: Implications for theory and practice. In K. A. Renninger, S. Hidi, & A. Krapp (Eds.), *The role of interest in learning and development* (pp. 361–395). Hillsdale, NJ: Erlbaum.

Renninger, K. A., & Hidi, S. (2000). Student interest and achievement: Developmental issues raised by a case study. In A. Wigfield & J. Eccles (Eds.), *The development of achievement motivation* (pp. 173–195). San Diego, CA: Academic Press.

Schiefele, U., Krapp, A., & Winteler, A. (1992). Interest as a predictor of academic achievement: A meta-analysis of research. In K. A. Renninger, S. Hidi, & A. Krapp (Eds.), *The role of interest in learning and development* (pp. 183–212). Hillsdale, NJ: Erlbaum.

Schunk, D. H. (1991). Self-efficacy and academic motivation. *Educational Psychologist, 26*, 207–231.

Shell, D., Murphy, C., & Bruning, R. (1989). Self-efficacy and outcome expectancy mechanisms in reading and writing achievement. *Journal of Educational Psychology, 81*, 91–100.

Skaalvik, E. (1997). Self-enhancing and self-defeating ego orientation: Relations with task and avoidance orientation, achievement, self-perceptions, and anxiety. *Journal of Educational Psychology, 89*, 71–81.

Stipek, D. (2001). Good instruction is motivating. In A. Wigfield and J. Eccles (Eds.), *The development of achievement motivation* (pp. 310–334). San Diego, CA: Academic Press.

Turner, J. C., Midgley, C., Meyer, D. K., Gheen, M., Anderman, E. M., Kang, Y., & Patrick, H. (2002). The classroom environment and students' reports of avoidance strategies in mathematics: A multimethod study. *Journal of Educational Psychology, 94*, 88–106.

Whang, P. A., & Hancock, G. R. (1994). Motivation and mathematics achievement: Comparisons between Asian-American and non-Asian students. *Contemporary Educational Psychology, 19*, 302–322.

Wigfield, A., & Eccles, J. (1992). The development of achievement task values: A theoretical analysis. *Developmental Review, 12*, 265–310.

Wolters, C., Yu, S., & Pintrich, P. R. (1996). The relation between goal orientation and students' motivational beliefs and self-regulated learning. *Learning and Individual Differences, 8*, 211–238.

Zusho, A., & Pintrich, P. R. (2003). A process-oriented approach to culture: Theoretical and methodological issues in the study of culture and motivation. In F. Salili & R. Hoosain (Eds.), *Teaching, learning, and student motivation in a multicultural context* (pp. 33–65). Greenwich, CT: Information Age.

Zusho, A., Pintrich, P. R., & Coppola, B. (2003). Skill and will: The role of motivation and cognition in the learning of college chemistry. *International Journal of Science Education, 25*, 1081–1094.

CHAPTER 8

JERRY WELLINGTON

PRACTICAL WORK AND THE AFFECTIVE DOMAIN: WHAT DO WE KNOW, WHAT SHOULD WE ASK, AND WHAT IS WORTH EXPLORING FURTHER?

INTRODUCTION: THE NEGLECTED DOMAIN?

In the late 1980s, John Head (1989) predicted that the focus of research interest in science education would change so that the "affective area ... will prove to be crucial in research and curriculum planning in the next decade." Although subsequent authors such as Claxton (1991) did write eloquently about students' views, emotions, and feelings about science, Head's prediction was not realised. Recently, with publications such as Simon (2000) and Osborne and Collins (2000), there has been some interesting survey work on the affective domain but it still receives less focus than its cognitive counterpart. This chapter reports on where we are with current thinking and research on the affective domain as it relates to practical work in school science, and offers a view on how we might go further. One of its aims is to reaffirm the importance of the affective (Woolnough, 2001).

PRACTICAL WORK AS EMBEDDED IN SCHOOL SCIENCE CULTURE

It required three sociologists, in 1988, to point out (from an outsider's perspective) some of the rituals and embedded practices that go on in school science in many so-called "developed countries." The three observers, wearing an almost "aliens from outer space" hat, called their article "In the beginning was the Bunsen" (Delamont et al., 1988) to show how the Bunsen Burner has become (for pupils and perhaps for teachers) the icon of school science. Since then, others have talked of the fact that science education and the school laboratory have become almost inseparable and certainly taken for granted as connected twins. Jenkins (1998, pp. 35–51) for example has written a persuasive history of the "schooling of laboratory science," or what I like to call the laboratorising of school science (Wellington, 1998, in the same book, p. 9). As Donnelly's study (Donnelly, 1995, p. 97) showed, teachers are often uncritical in their approach to practical work. One teacher, who seemed to be typical, was quoted as saying "it's what science is all about really, is getting on with some experiments. Science is a practical subject ... you know, end of story I think." Only recently have critical thinkers in science education begun to question the taken-for-granted link between learning science and the school science lab, using terms such as the Gordian knot (Osborne, 1998, pp. 156–173).

Steve Alsop (ed.) Beyond Cartesian Dualism, 99–109.

THE SCEPTICS

One of the ironies worldwide is that those countries where practical work is well established and most prevalent are the contexts in which it is most criticised—those doing it question it, those who strive to do more are less sceptical (Leach, 1999; Leach & Paulsen, 1999a, 1999b). Practical work has been present in British schools for around 150 years (Gee & Clackson, 1992) yet critical comment on it is relatively recent, despite the fact that vast sums of money and time have been, and are, devoted to it.

Critics of the conduct and purpose of practical work include Driver (1975) and Wellington (1981) who pointed out the limitations and "game playing" involved in the discovery approach; then later in the 1980s, Woolnough and Alsop (1985) questioned the link between practical activity and scientific knowledge and this has continued (e.g. Millar, 1998); in the 1990s, Hodson in many important publications (starting with Hodson, 1990) questioned the value of practical work, either for developing transferable skills or for providing authentic science experiences (see also Woolnough, 1998); then in the middle of the same decade, Donnelly and colleagues (1996) acted as a voice for many who questioned the imposition of assessed investigations on school science teachers. At the turn of the century, Watson (2000) summed up the views of many science educators and classroom teachers by saying: *Does it work? Is it worth the investment? Can it be used more effectively?*

TEACHERS' BELIEFS

Teachers have a range of beliefs about practical work in school science and the justifications for doing it. It may appear strange, but far more exploration has been made of teachers' views on and attitudes to practical work than those of their students. We cannot explore these in full here, but past studies show that teachers commonly appear to hold the view that practical work can act as a motivator for pupils, especially those in the early years of secondary school. Kerr's extensive study (Kerr, 1963) for example showed that teachers rated the aim of "to arouse and maintain interest in science" as the most important objective for practical work in years 7–9. More recent studies have confirmed Kerr's work during 1960s, showing in addition that teachers often tend to make unquestioned assumptions about the role of practical work in motivating and in enhancing learning (see for example, Hodson, 1990,1992, 1993; Hofstein, 1988; Welzel et al., 1998). Teachers do not always have a clear vision of what purpose a particular practical might serve, but they get the pupils to do it anyway (Harlen, 1999)—perhaps in the hope that it acts as a break from "theory" or simply a change or diversion in the lesson (Wellington, 1998).

PRACTICAL WORK AND STUDENT MOTIVATION

Thus the literature shows that studies of teachers' perceptions indicate a widespread view that they believe practical work to be "motivating." But the questions for this chapter are can teachers' assumptions be justified? Is there any evidence from studies of pupils' views or from broader research into the nature and efficacy of practical work that might support their views?

The following questions need to be asked:

- Does practical work motivate all pupils, all pupils equally, or some more than others?
- What does it motivate pupils to do?
- And exactly *why* does it motivate pupils?

One of the things I noticed in producing Wellington (Ed. 1998) is that there is not a vast wealth of empirical work to draw upon in answering these questions. Pupils and students have rarely been asked straightforward questions such as do they enjoy practical work? Does it motivate them? (We also need to ask in any attribution of motivation: "Motivated to do what?") And finally, what do they learn from it? Denny and Chennell (1986) did explore "pupils' views and feelings" about practical work in school science by asking them to complete a letter-writing task or a drawing exercise. Their study seemed to reveal that pupils often saw practicals as confirmatory rather than investigatory exercises (things may have changed since then of course) and they saw them as a teacher device to reduce boredom. However, pupils saw them as a source of enjoyment. (We need to ask the further question: why did they enjoy them?)

In an earlier study, Moreira (1980) found that, in physics practical work, students felt that it involved little more than following instructions. Later, in a study of investigational work, Watson and Wood-Robinson (1998) found that students were not aware of why they were doing them. Other studies (for example, Lunetta, 1998; and Woolnough, 1991 & 1998) have shown that for most students laboratory work involves manipulating equipment rather than manipulating ideas—cognitive engagement is rare (Watson, 1994).

Recent studies of pupils' general attitudes to science are also worth referring to. Osborne and Collins (2000) found that pupils see science as valuable and important, but also difficult and not of great intrinsic interest. They expressed a greater interest in work that involves "experimenting" and investigating. Simon (2000) reviewed research on pupils' attitudes and surmised that positive attitudes to science peak at the age of 11 and decline thereafter. Further back, Myers and Fout (1992) suggested that positive attitudes to science are improved by the use of hands-on activities involving group collaboration. Hofstein (1988) also claimed to show that laboratory work does develop positive attitudes and interest in science. Earlier, however, Head (1982) suggested that not all students are motivated by practical work and we should examine the reactions of students to various types of practical activity, looking into the sex, age, and ability differences in their responses (see also Kempa & Diaz, 1990 and Murphy, 1991).

Several authors (as far back as Kreitler & Kreitler, 1974) have, quite rightly, asked the question: exactly why are many pupils motivated by practical work?

One of the more useful concepts for considering the science laboratory was presented by McComas (1997) who suggested that we view it from an ecological perspective. The environment would then be divided into *physical factors* and *personal factors*. The latter would include not only the students' prior knowledge and ability but also their attitudes and motivation. In the same vein, White (1988) argued that the context of a student's learning is vital and this is determined by, not least, the "social dimension." Others (such as Giddings et al., 1990 and Fraser & Wubbels, 1995) have made similar pleas for a greater focus on the learning environment of the school laboratory. In my view, the ecological perspective is a valuable one in studying the affective domain.

With these discussions in mind, in the next section I present some recent data on the views and attitudes of students, some of my own and some from others.

RECENT EMPIRICAL DATA ON PRACTICAL WORK AND THE AFFECTIVE DOMAIN

In carrying out work for the QCA in Britain in 1998 (see Nott and Wellington, 1999 and Nott et al., 1998) I was able to interview large numbers of school pupils between the ages of 11 and 18. We were unable to publish all of the interview data due to lack of space. I have included some of that unpublished data here from my discussions with a range of school pupils.

Pupils in early years of secondary schooling were largely positive about practical work, often using terms like "it was fun" or "it was a break" (from "theory"—this raises a question about the connection with cognition, which we return to later). A minority of pupils felt that it reinforced their theoretical learning, but others said that practical confused them if the results were not as expected and when they did not conform to theory.

Several pupils raised the issue of group work. For many this is an enjoyable experience but most said they disliked working in groups with more then two other pupils. Even at this age (11–14) pupils could see the difference between "ordinary practicals" and "investigations." They were able to see that some "experiments" (as they call all practical activity) were only done to prove a point and not really to "investigate."

Older students (year 12, aged 17 and above) who were continuing with science and therefore well disposed towards it were interviewed in the study. They had strong views about practical work. The sentiment again expressed (similar to the view of younger pupils that practical work is a diversion or a break) was often that "practical work is better than theory," showing that the divide persists through secondary schooling. Several expressed the importance of "seeing things happening" and felt that it helped you to remember things and perhaps to understand them. Others, an equal number, commented on negative experiences:

> Practicals can confuse you if they don't work
> You had to go on and do it even though you knew what would happen
> I had to write a conclusion for an experiment I didn't even do
> Something always went wrong with Physics practicals

These older pupils were also able to look back on their experiences from when they were 11–16 years old and see the repetition that had occurred, for example

> I've done the huddling penguins experiment 80,000 times
> We saw magnesium ribbon burnt 3 times in key stages 3 and 4

They were rather more sceptical (even cynical) towards the assessed investigational work they had carried out up until year 11 (age 16) with comments such as

> It's more about getting a good mark than learning science
> Teachers try to get you the best mark they can so they can get higher in the league tables. It's their job
> You fiddled quite a bit . . . if a result wasn't in line you changed it

On the other hand, one of the more astute students commented that

> You've got to have an anomalous result, even if you didn't have one

(This student used the word anomalous herself, sometimes they used the word "dodgy," but they all knew that to get good marks in the evaluation phase of an investigation you needed a few dodgy results: see also Toplis, 2003 and Keiler & Woolnough, 2002).

The data above were collected in 1998 and 1999. For this chapter, I decided that I needed some more recent data from small groups of pupils, focusing on their attitudes and feelings towards practical work. The interviews reported below were carried out by a teacher I have been working with (Toplis, 2003, PhD thesis) who knew the pupils in his school and was perhaps more able than me to have a frank conversation with them. Two groups of pupils were involved, one in year 8 and one in year 10.

We posed five main questions: What are your feelings about practical work in science? What do you like about it? What do you dislike about it? Does it help you to learn science, and if so how? Does it make you keener and more motivated towards science? This acted as a loose framework for the interviews and led to some interesting focus group discussions.

FEELINGS

Views expressed were that practical work is fun, exciting, "better than writing," it "gives you a break," "some relaxation," "it's different":

> It's fun to do something, not just writing
> It's nice to see what happens
> It's easier than reading about it
> You can talk about what you're doing as well as doing what you're doing—you're all working together
> It's more interesting than reading or listening to teachers
> It makes you want to do well in science

Some value practical work because it is unique to science: "We don't do things like it at home, or in any other lesson."

Aspects they disliked were

> When I don't understand what to do or why
> If you don't know exactly what you're supposed to be doing and why you're doing it
> Sheets of instructions with hard words on them
> Setting up at first, then clearing away and tidying up
> Touching dangerous chemicals when we don't know exactly what's wrong with them

They also commented on a dislike of being rushed and having to fit their practical work into a time slot.

HOW DOES IT HELP IN LEARNING SCIENCE?

One pupil in our sample mentioned the idea that skills can be acquired:

> You can pick up skills when you're doing it yourself

The idea that practical work provides a visual picture came up several times:

> Although we know what's going to happen, you can't quite picture it
> It gives you a visual picture
> When you're revising you can visualise yourself doing things

In general, the idea of practical work being an *aide-memoire* came up commonly, especially when "you don't take stuff in by reading it." Pupils tend to remember episodes and things that are "out of the ordinary."

GROUP WORK

Pupils were very forthcoming with comments about group work. They value it, commenting that

> You can work with your friends
> You don't get much chance in other lessons
> You can help each other out, there's no need to get stuck

But they also had certain reservations:

> It depends who is in the group
> With too many people in the group, you can't get a look in
> Sometimes, one person wants to do everything
> If you're all different abilities it can be a problem
> There's often one person who does nothing and tries to take all the credit
> If things go wrong, group members can fall out

One pupil, ahead of his time perhaps, observed that when the class is doing practical work in groups it can "help the teacher to pinpoint people with special needs."

A similar study, but on a larger scale, was carried out by Spriggs (2003) as part of his work for a doctorate. He conducted 14 in-depth focus group discussions with boys and girls from years 7 to 9 (ages 11–14) in his own school. The groups were asked to discuss three main areas: their reasons for liking practical work; their reasons for believing practical work to be useful; and their views on the learning that can be derived from doing practical work. The pupils' perceptions related to both teacher demonstrations and class/ group work.

Whilst admitting the limitations of the focus group as a research tool (see Wellington, 2000, pp. 124–127) my view is that Spriggs has provided some useful, new insights into pupils' feelings about practical work and their views on why it does (and does not) help them learn science. I cannot do the study justice in the space afforded here, but some of the main points are summarised below.

REASONS FOR LIKING PRACTICAL WORK

Pupils' reasons for liking practical work (PW) relate well to the data presented above. Spriggs distinguished six areas from his data: fun/interest, fascination, danger, surprise, achievement/satisfaction, and independence. The most common reason had to do with the "fun" of doing it, especially when they work in small groups, doing their own "experiment." Pupils also mentioned the interest factor, either in seeing, hearing, or smelling things; and they commented on the interest aroused by actually "doing it," instead of reading or writing about it. The fascination factor is a similar reason, pupils often citing the sights, sounds, and occasional smell (and even taste) involved in PW. Some groups, for example in the use of Bunsens or acids, cited a sense of danger as one of the attractions of PW. But for a minority, the danger factor was a reason for disliking PW. The element of surprise was a motivating factor for many, but only if they did *not* know what was likely to happen. For many, the sense of achievement and satisfaction is a motivator, when they actually make something work or do something for themselves, or "get something out of it." Finally, the feeling of being independent or autonomous and "being in charge," instead of just watching or listening, was valued.

REASONS FOR BELIEVING PRACTICAL WORK TO BE USEFUL

In this area, Spriggs divided the themes emerging from the focus groups into four factors: the beneficial for learning factor, the effortless factor, the relaxation factor, and the socialization factor.

Firstly, pupils feel that PW can help their learning, principally because they can get a "picture image" (one pupil's words) inside their head—by watching something happen, they will not forget it (c.f. the data above). This was said to be easier and involve less effort than listening or writing from the board—one said that it is more likely to go into "the memory bank inside your head" if you have seen it rather than heard or read about it. The relaxation factor related to their view that PW is less formal and they feel more relaxed (it feels less like work), so that they find it easier and more fun to learn. Finally, many pupils felt that by working in groups they could chat about things, learn from their peers, and help each other. However, as in the data above, several students felt that socialization and group work could be a problem, especially in groups of more than three or if group members did not "get on" or behave responsibly.

THE LEARNING THAT DERIVES FROM PRACTICAL WORK

The learning perceived to result from PW was divided into two categories: the reality factor and the procedures factor. On the other hand, pupils also commented on three areas where they felt that PW did not really enhance their learning: practical skills, scientific processes, and thinking about theory. We start with the perceived learning benefits.

Firstly, pupils felt that first-hand, real experiences of phenomena were valuable for learning. Again they felt that it helped them remember more, but also that it helped them believe it (as opposed, presumably to seeing it on a computer or TV screen). PW was seen as making things come alive and real. They also felt some "responsibility" for it, if they did it themselves. Some even mentioned the value of doing it or seeing it happen when things went wrong, especially if they had the opportunity to do it again. The second factor, related to actually doing it, was the feeling that they had carried out and learned about the right procedures and could remember how to do things as a result again, especially if they could repeat an "experiment."

There were three areas where pupils were more sceptical about their learning gains. Firstly, learning about scientific processes and being a scientist: pupils knew that they were not really doing the "important things" that real scientists do, partly because they are doing it in order to learn. They know that people have done these "experiments" before and what will happen is well known. They are not discovering anything but learning instead. Secondly, they were not always convinced that the skills they were learning were generic or transferable to contexts outside the school laboratory. Some felt that skills such as measuring or following instructions could be learnt just as well in other contexts e.g. food technology. Finally, and perhaps most importantly for one of the key themes of this chapter, pupils were not convinced that practical work helped them to understand scientific ideas or "theory." One commented that "it's practical work, no thinking, just watching and doing . . . you really don't need to think." It requires limited mental effort. This links to earlier comments that PW is related to relaxation, a break, a diversion, entertainment rather than thinking.

MAKING SENSE OF THE EMPIRICAL DATA: LINKING PRACTICAL WORK, AFFECT, AND THE COGNITIVE DOMAIN

The data presented above are derived from small samples, but they do provide an initial basis to help conceptualise the field in the area of practical work and the affective domain. They, and earlier studies cited in this chapter, show that we should accept that practical work does play an important and positive role in improving some students' attitudes towards science, their keenness to do science, and perhaps even their self-esteem, i.e. self-belief that they can actually do some science. However, the key questions outlined at the start of this chapter still need to be fully addressed: Does practical work motivate all pupils (and if so, why?), what does it motivate them to do and how can we build on this? Does practical work have any greater influence on the affective domain, e.g. attitude to science, than other teaching and learning strategies? There is still enormous scope for, and obvious value in, further research on the area of practical work and affect. Future research could usefully employ the ecological perspective suggested by McComas in 1997. This would involve a holistic study of the *physical factors* (the laboratory itself, its paraphernalia, the resources, the layout, the logistics of practical work) and the *personal factors* (the feelings of the pupils, their own identities and backgrounds, interactions between peers, interactions between teachers, students, and laboratory technicians.). The laboratory could be viewed as a kind of habitat in which certain behaviours and rituals take place (as they do in any learning environment). The ecology of learning would then be the main focus, perhaps adopting many of the concepts of ecological theory to understand the science-learning environment.

In taking an ecological perspective, we might also address a recurring and even more complex question for science education: How does practical work and the affective domain relate to the *cognitive* domain in science learning? Clearly, and this is almost a tautology, pupils will only get better at doing practical work by doing practical work (provided, as a famous golfer once said, they don't keep practising their faults i.e. good teacher intervention and coaching is an essential). But doing practical work is only likely to improve their procedural skills and perhaps their procedural thinking (or at the base level, their ability to follow instructions and to manipulate apparatus). Is it likely to increase their knowledge of science or at a higher level, their understanding of science and scientific theory? Similarly, will it improve their knowledge, attitude to and understanding of the nature of science and scientific enquiry, unless the practical work attempts to mimic authentic science *and is successful in achieving this aim?* (see Woolnough, 1998; Hodson, 1985, 1993)

We need to distinguish between knowledge that and knowledge how (first made by Gilbert Ryle in 1949) and a third category of knowledge in school science that I have labelled "knowledge why" (Wellington, 1989, p. 11). Engaging in practical activities may well improve pupils' knowledge of what happens (knowledge that) and even improve their ability to remember it, i.e. factual recall. It may well improve their knowledge of how things happen, and their knowledge of how to do things (Ryle's original meaning of "knowledge how"). But does it have any impact on their knowledge of why things happen, i.e. the essential point of science and scientific explanation? Does practical work impinge upon anything other than phenomenal knowledge?

Evidence would suggest that it does not. The essential bridge that needs to be built between the world of experiences (the phenomenal) and the world of explanations (the conceptual or theoretical) is often never constructed. It is certainly not constructed simply by *doing things.* Several people have called these two areas, between which their seems to be a chasm more often than a bridge, the "hands-on" and the "minds-on" (see for example Woolnough and Millar's two chapters in Wellington, 1998). A key role for the teacher is to help to build the bridge or scaffold between the world of things, objects, events, and phenomena, and the world of ideas, theories, and abstractions (the scientific story). This can only be built by the use of words, language, discussion, and teaching. The gap between *knowledge that* and *knowledge why* will not be built by unguided "discovery" or unsupported and unscaffolded activity amongst groups of children. For children to learn *science* (and be truly motivated by it) as opposed to "blindly" observing phenomena or carrying out procedures, the teacher has to be more than a classroom supervisor or facilitator, no matter how excited students are by the busy-ness and the business of doing practical work.

ACKNOWLEDGEMENTS

I would like to thank Rob Toplis, one of my former PhD students, for the help he provided by interviewing pupils in his own school and for the stimulating meetings we had in connection with school science practical work during the course of his research.

REFERENCES

Claxton, G. (1991). *Educating the inquiring mind: The challenge for school science.* Hemel Hempstead: Harvester Wheatsheaf.

Delamont, S., Beynon, J., & Atkinson, P. (1988). In the beginning was the Bunsen: The foundations of secondary school science. *Qualitative Studies in Education, 1*(4), 315–328.

Denny, M., & Chennell, F. (1986). Science practicals: What do pupils think? *European Journal of Science Education, 8*(3), 325–336.

Donnelly, J. (1995). Curriculum development in science: The lessons of Sc1. *School Science Review, 76*, 95–103.

Donnelly, J., Buchan, A., Jenkins, E., Laws, P., & Welford, G. (1996). *Investigations by order.* Nafferton: Studies in Education.

Driver, R. (1975). The name of the game. *School Science Review, 56*(197), 800–804.

Fraser, B. J., & Wubbels, T. (1995). Classroom learning environments. In B. J. Fraser & H. J. Walberg (Eds.), *Improving science education.* Chicago: National Society for the Study of Education.

Gee, B., & Clackson, S. G. (1992). The origin of practical work in the English school science curriculum. *School Science Review, 73*, 79–83.

Giddings, G. J., Hofstein, A., & Lunetta, V. (1990). Assessment and evaluation in the science laboratory. In B.E. Woolnough (Ed.), *Practical science.* Milton Keynes, United Kingdom: Open University Press.

Harlen, W. (1999). *Effective teaching of science* Edinburgh: SCRE.

Head, J. (1982). What can psychology contribute to science education? *School Science Review, 63*, 631–642.

Head, J. (1989). The affective constraints on learning science. In P. Adey, J. Bliss, & M. Shayer (Eds.), *Adolescent development and school science* (pp. 162–169). London: Falmer.

Hodson, D. (1985). Philosophy of science, science and science education. *Studies in Science Education, 12*, 25–57.

Hodson, D. (1990). A critical look at practical work in school science. *School Science Review, 70*, 33–40.

Hodson, D. (1992). Redefining and reorienting practical work in school science. *School Science Review, 73*, 65–78.

Hodson, D. (1993). Re-thinking old ways: Towards a more critical approach to practical work in school science. *Studies in Science Education, 22*, 85–142.

Hofstein, A. (1988). Practical work and science education. In P. Fensham (Ed.), *Development and dilemmas in science education* (pp. 189–218). London: Falmer.

Jenkins, E.W. (1998). The schooling of laboratory science. In J. Wellington (Ed.), *Practical work in school science: Which way now* (pp. 35–51)? London: Routledge.

Keiler, L. S., & Woolnough, B. E. (2002). Practical work in school science: The dominance of assessment. *School Science Review, 83*(304), 83–88.

Kempa, R., & Martin Diaz, M. (1990). Motivational traits and preferences for different instructional modes in science. *International Journal of Science Education, 12*(2), 195–203, 205–216.

Kerr, J. (1963). *Practical work in school science.* Leicester: Leicester University Press.

Kreitler, H., & Kreitler, S. (1974). The role of the experiment in science education. *Instructional Science, 3*, 75–88.

Leach, J. (1999). Introduction to practical work and learning about science. In J. Leach & A. C. Paulsen (Eds.), *Practical work in science education:Recent research studies* (pp. 115–117). Frederiksberg: Roskilde University Press/Dordrecht: Kluwer.

Leach, J., & Paulsen, A. C. (Eds.). (1999a). *Practical work in science education: Recent research studies.* Roskilde University Press.

Leach, J., & Paulsen, A. (1999b). Introduction: How this book came about, and its contribution to the literature. In J. Leach. & A. Paulsen (Eds.), *Practical work in science education: Recent research studies* (pp. 7–13). Frederiksberg: Roskilde University Press/Dordrecht: Kluwer.

Lunetta, V. N. (1998). The school science laboratory: Historical perspectives and contexts for contemporary teaching. In B. J. Fraser & K. G. Tobin (Eds.), *International handbook of science education.* Dordrecht: Kluwer Academic.

McComas, W. F. (1997). The laboratory environment: An ecological perspective. *Science Education International, 8*(2), 12–16.

Millar, R. (1998). Rhetoric and reality: What practical work in science education is really for? In J. Wellington (Ed.), *Practical work in school science: Which way now* (pp. 16–33)? London: Routledge.

Moreira, M. A. (1980). A non-traditional approach to the evaluation of laboratory instruction in general physics courses. *European Journal of Science Education, 2*, 441–448.

Murphy, P. (1991). Gender differences in pupils' reactions to practical work. In B. E. Woolnough (Ed.), *Practical science.* Milton Keynes, United Kingdom: Open University Press.

Myers, R. E., & Fouts, J. T. (1992). A cluster analysis of high school science classroom environments and attitude toward science. *Journal of Research in Science Teaching, 29*(9), 929–937.

Nott, M., Peacock, G., Smith, R., Wardle, J., Wellington, J., & Wilson, P. (1998). *Investigations into KS3 and KS4 science. A report prepared for the Qualifications and Curriculum Authority.* Projects 10905 and 10906. Sheffield: Sheffield Hallam University.

Nott, M., & Wellington, J. (1999). The state we're in: Issues in key stage 3 and 4 science. *School Science Review, 81*(294), 13–18.

Osborne, J. (1998). Science education without a laboratory? In J. Wellington (Ed.), *Practical work in school science: Which way now* (pp. 156–174)? London: Routledge.

Osborne, J., & Collins, S. (2000). *Pupils' and parents' views of the school science curriculum.* London: King's College. Available at: <kcl.ac.uk/depsta/education/publications/Pupils_Report.pdf> (Version current at November 18, 2004).

Ryle, G. (1949). *The concept of mind.* London: Hutchinson.

Simon, S. (2000). Students' attitudes towards science. In M. Monk & J. Osborne (Eds.), *Good practice in science teaching* (pp. 104–121). Buckingham: Open University Press.

Spriggs, G. J. (2003). *How do children's conceptions and contextual preferences affect what they do and learn in practical science classes?* Unpublished EdD thesis, Institute of Education, University of London, London.

Toplis, R. (2003). Key stage four pupils' evaluations of science investigations. Unpublished PhD thesis, University of Sheffield, Sheffield.

Watson, J. R. (1994). Students' engagement in practical problem solving: A case study. *International Journal of Science Education, 16*(1), 27–43.

Watson, R. (2000). The role of practical work. In M. Monk & J. Osborne (Eds.), *Good practice in science teaching* (pp. 57–72). Buckingham: Open University Press.

Watson, R., & Wood-Robinson, V. (1998). Learning to investigate. In M. Ratcliffe (Ed.), *ASE guide to secondary science education.* Hatfield: ASE.

Wellington, J. (1981). What's supposed to happen, Sir? Some problems with discovery learning. *School Science Review, 63*, 167–173.

Wellington, J. (Ed.). (1989). *Skills and processes in science education.* London: Routledge.

Wellington, J. (Ed.). (1998). *Practical work in school science: Which way now?* London: Routledge.

Wellington, J. (2000). *Educational research: Contemporary issues and practical approaches.* London: Continuum.

Welzel, M., Haller, K., Niedderer, H., & von Aufschnaiter, S. (1998). *Teachers' objectives for labwork.* Working Paper 6 Labwork in Science Education Project. Available at: <www.physik.uni-bremen.de/physics.education/niedderer/projects/labwork/papers.html> (Version current at November 18, 2004).

White, R. T. (1988). *Learning science.* Oxford: Blackwell.

Woolnough, B. (1991). *Practical science.* Buckingham: Open University Press.

Woolnough, B. E. (1998). Authentic science in schools, to develop personal knowledge. In J. Wellington (Ed.), *Practical work in school science: Which way now* (pp. 109–125)? London: Routledge.

Woolnough, B. E. (2001). Of 'knowing science' and of 'doing science': A reaffirmation of the tacit and the affective in science and science education. *Canadian Journal of Science, Mathematics and Technology Education, 1*(3), 255–270.

Woolnough, B., & Alsop, T. (1985). *Practical work in science.* Cambridge: Cambridge University Press.

CHAPTER 9

LYNN D. DIERKING

MUSEUMS, AFFECT, AND COGNITION: THE VIEW FROM ANOTHER WINDOW

INTRODUCTION

In 1994 the Institute for Learning Innovation hosted a National Science Foundation-funded conference in Annapolis, Maryland, to discuss the creation of a research agenda that would investigate the long-term impact of experiences in museums and science centers. As a warm-up exercise on the first day, participants were invited to share a memorable museum experience of their own. Most participants represented science museums, science centers, aquariums, and zoos but despite the preponderance of science museum professionals, the memorable experiences that participants described were not about learning some fascinating *factoid* of science. Instead, the experiences shared often included a close interaction with an object or idea of personal significance to the participant, which was described in a rich and emotional way (I can still remember the vivid description of one person, very interested in the history of astronomy. He described a moment of epiphany gazing at an ancient mariner's sexton displayed at the Greenwich Museum in Greenwich, England, with such emotion that it brought tears to *my* eyes).

Although this activity was designed as a warm-up, a way to begin the conference discussion, it became a focal point of the meeting, a conversation that was returned to throughout the meeting, as the group struggled to craft recommendations for a long-term research agenda to document learning in and from museums. I think the group returned to these powerful and moving stories, not because of their details, but because there were some central truths about learning contained within them. For me personally, it was a defining moment, an experience that I have found myself continuing to think about over the last decade and which has shaped my research and ideas significantly. The individual stories shared encapsulate for me both the wonder and challenge of understanding and documenting science learning, certainly in the arena of museums and science centers, but in other settings as well.

Recent and emerging research about the nature of learning, encapsulated well in Mike Watts' chapter and others in this book, suggests that learning is always highly personal, multifaceted and strongly influenced by setting and emotion. However, as so many of the chapters in this book also attest, demonstrating the role that affect plays in the learning of science, particularly in classrooms, is difficult and time consuming; such investigations are rare and still in their infancy (fortunately this book gathers a number of them together in a very comprehensive and accessible manner).

Steve Alsop (ed.) Beyond Cartesian Dualism, 111–122.

However, there is another perspective. Is it possible that free-choice science learning arenas such as museums, science centers, zoos, aquariums, the Internet, and the family are actually better settings in which to investigate the relationships between affect and cognition in learning? I think so, which is why I have been drawn to these settings as places to investigate learning, particularly learning guided by the needs, interests, and motivations of learners themselves. I believe that the unique qualities of these settings (their interdisciplinary and multisensory presentation of people, events, ideas, and objects, their ability to not only accommodate but encourage and sometimes even *require* social interaction and the opportunities afforded for choice and control by the learner) allow a researcher a window with a different, but complementary and perhaps even better view, for investigating the role that affect plays in facilitating science learning.

In this chapter I plan to expand upon these ideas, by building a case for this argument and sharing some of the empirical studies that colleagues and I have conducted in this area. I have three goals: (1) To briefly describe the role that affect plays in learning, utilizing evidence emerging from new brain research; (2) To discuss the important role that affect plays in science learning in and from museums; and (3) To share some of the studies which have attempted to demonstrate the role of affect in such learning. Finally I will suggest the need for a more integrated understanding of emotion and cognition across all aspects of the science-learning infrastructure (schools, universities, museums, science centers, etc.).

AFFECT AND LEARNING

"When she got to the river's edge, she found that a large boat was travelling down the river and the lift bridge was up. She waited while the boat passed and the bridge came down. While waiting she watched the mechanism of the bridge, and realized excitedly that she understood the way that the counterweights and gears were making the huge mass of iron, steel, and roadway go up and down so easily. She also realized that she had a whole vocabulary to describe this that she didn't realize she knew. Then she recalled that she had learned all this from an exhibit on bridges at the Cleveland Children's Museum, which she had visited the year before with her young grandson."

This brief but intriguing description shared by a visitor illustrates the "view from a different window." This woman had not participated in a formal physics course in which bridges were discussed nor were there plans for a test at the end of her experience. Although well educated, serving as an Assistant Professor of Speech & Hearing at a university, her areas of interest are language and reading development, not science. She even describes herself as a science novice. However, she was able to richly describe how counterweights and gears act together to move a huge mass of iron, steel, and roadway up and down easily, using technical vocabulary that she did not realize she knew. How is this possible?

This woman's experience at the bridge demonstrates three important concepts central to the nature of the learning process and accommodated well in free-choice settings: Learning is strongly influenced by (1) emotion and other affective factors; (2) motivation; and (3) closely related to the previous two principles; learning is facilitated by personal interest. In addition, "new" knowledge is constructed from a foundation of prior experience and knowledge and is expressed within appropriate contexts.

EMOTION AND OTHER AFFECTIVE FACTORS

Most human learning is self-motivated, emotionally satisfying, and very personally rewarding. Humans learn when they are in supporting environments, when meaningful activities are engaged in, when they are freed from anxiety, fear, and other negative mental states,when individuals have choice and control over their learning, and when the challenges of the task meet the person's skills (Covington, 1992; Csikzentmihalyi, 1990a, 1990b; Deci 1992; Deci et al. 1981; Deiner & Dweck, 1980; Maehr, 1984; McCombs, 1991; Paris, 1997; Paris & Cross, 1983; Pintrich & DeGroot, 1990). Within this context, it is easy to imagine the woman described earlier, readily learning about the role of gears and counterweights in the functioning of bridges as she plays with her grandchild at a children's museum.

Learning as a process/product evolved long before there was language or mathematics, in fact, before there were humans, primates, or other mammals at all. It is profoundly important to appreciate the long evolutionary history of human learning; learning is not just a recent cultural overlay unique to "modern humans." A very important, and relatively unappreciated byproduct of this long evolutionary history is the feedback loops that exist between emotional and cognitive processes. In large part these feedback loops are mediated by one of the oldest parts of the brain, the area known as the limbic system. All incoming sensory information is given an initial screening for meaningfulness and personal relevance by structures in the limbic system. This process both determines *what* is worth attending to and remembering and *how* something is remembered. Research in the last quarter century has shown that learning cannot be separated in the Cartesian sense between rational thought and emotion, nor neatly divided into cognitive (facts and concepts), affective (feelings, attitudes, and emotions), and psychomotor (skills and behaviours) functions as many psychologists and educators have attempted to do for nearly a half century (Rose, 1993; Sylwester, 1995). All learning, even of the most logical topic, involves emotion, just as emotions virtually always involve cognition (Damasio, 1994; Piaget, 1981).

By virtue of its journey through the limbic system, every memory comes with an emotional "stamp" attached to it (Damasio, 1994). The stronger the emotional "value," the more likely sensory information is to pass this initial inspection and be admitted into memory; and interestingly, pleasant experiences are strongly favored over unpleasant ones (Damasio, 1994; Sylwester, 1995;); implicit in this notion of learning is that learning is a pleasurable experience, one that learners seek out. Evolution has thus insured a dependency between learning and survival by making the process of acquiring and storing information both very thorough and, for the most part by virtue of its relationship to the limbic system, more often than not an intrinsically pleasurable and rewarding experience (Csikzentmihalyi, & Hermanson, 1995). As theorized by neuroscientist Gerald Edelman, learning is a whole body experience, involving the emotions, the senses, the physical as well as the mental (Rosenfield, 1990). One place where this connection is seen powerfully is in the well-designed free-choice learning setting, multisensory settings that facilitate emotional, whole body learning experiences effectively.

MOTIVATION

There is another key idea about learning that is central to this thesis. Almost 50 years ago, psychologists realized that a basic dichotomy existed in learning; people either learned when they felt they wanted to or they learned because they felt they had to (Harlow, 1954). The outcomes of learning, it seemed, differed significantly depending upon which of these two conditions, or *motivations* existed. The terms used to distinguish between these two types of motivation were *intrinsic* versus *extrinsic* motivation (Csikzentmihalyi and Nakamura, 1989). Action is extrinsically motivated when the anticipated benefits are external to the activity. By contrast, intrinsic motivation means that the performance of an action is done for its own sake, even in the absence of some external reward. Adult participation in evening arts and crafts, exercise and relaxation classes, visiting a museum or theater while on vacation, and playing sports and games after school are examples of intrinsically motivated activities (Deci & Ryan, 1985; deCharms, 1992; White, 1959).

The affect and behaviors observed in museums and other engaging free-choice settings closely resemble the descriptions of learning recorded by psychologists investigating intrinsically motivated learning. Originally studied and described by psychologist Mihalyi Csikzentmihalyi, and confirmed by a wide range of other investigators, people appear to exhibit a common set of behaviors and outcomes when engaged in free-choice tasks for which extrinsic rewards are absent (Clifford, 1991; Csikzentmihalyi & Hermanson, 1995; Schunk, 1989). Chess players, rock climbers, dancers, painters, and musicians use similar explanations when describing the attraction of the activities they enjoy doing. They stress the fact that what keeps them involved in these demanding activities is an inherent quality, something Csikzentmihalyi calls the *flow experience*, because it is generally described as a state of mind that is spontaneous, almost automatic, like the flow of a strong current (Csikzentmihalyi & Hermanson, 1995).

A general characteristic of intrinsically motivated activities that produce flow is that they have clear goals and appropriate rules. Flow activities also usually provide immediate and unambiguous feedback and according to Csikzentmihalyi, this constant accountability is a major reason one gets so completely immersed in a flow activity. Another universally mentioned characteristic of flow experiences is that they tend to occur when the opportunities for action in a situation are in balance with the person's abilities. If the challenges are greater than the skill levels, anxiety results; if skills are greater than challenges, the result is boredom (Csikzentmihalyi & Hermanson, 1995; Rohrkemper & Corno, 1988). This phenomenon appears to hold across a wide array of skills, including physical, mental, artistic, and musical talents.

Successful museum exhibitions, performances, films, television programs, and Web sites share these qualities as well as one other fundamental component of motivation, the element of control (Paris, 1997). In free-choice learning situations, the learner can self-select the challenge they wish, rather than have it imposed upon them. In a museum this is evidenced by the high degree of self-selection that visitors exercise over which exhibits to view and/or utilize. No visitor views/utilizes all exhibits in a museum, but virtually every visitor views/utilizes some (Falk & Dierking, 1992). Generally, the ones selected are ones that interest the visitor, as well as provide appropriate levels of intellectual, physical, and emotional challenge.

INTEREST

Many psychologists, and educators, treat motivation as a vague, everyday term, of which interest is a component. Psychologist Ulrich Schiefele has pointed out that motivation, in general, and interest, in particular, are complex, multidimensional states; even the construct of intrinsic motivation described above is usually subject or topic specific, rather than a generalized quality (Schiefele, 1991). Thus, when I use the term *interest* I am not merely referring to what someone likes or dislikes. Rather, I refer to a psychological construct that includes attention, persistence in a task, and continued curiosity, all factors important to an understanding of what might motivate someone to learn in a museum—to become fully engaged in a museum exhibition, program, or event (Dierking & Pollock, 1998; Hidi, 1990).

People are bombarded with stimulation all the time. The human brain is designed to sift through this abundance of information to selectively determine what to attend to and what to ignore. One filter for this selection process is *interest*. If we had no interests, our senses would be deluged with information and total mental chaos would result. As pioneering psychologist William James stated more than a hundred years ago, "without selective interest, experience would be utter chaos." [James (1890/1950), p. 237]

What determines interests includes a range of variables, some of which are universal, some the result of individual experiences, and others personal history. When people like something, they attribute positive feelings and values to it; the result is a high probability that they will choose to follow up on that interest with action (Pintrich & DeGroot, 1990). A wide range of investigators have remarked on the presence and influence of "interest" in the learning and behaviour that occurs in free-choice learning settings like museums (Ramey-Gassert et al., 1994). Successful museum exhibitions, performances, films, television programs, and Web sites are excellent media in which visitors can pursue their own personal interests because there is sufficient breadth and depth to permit the learner to self-select the experience they wish to have, based on their prior knowledge, experience, and interests.

AFFECT AND SCIENCE LEARNING IN AND FROM MUSEUMS

One action that can, and for many people does flow from interest in science and technology, is the decision to attend a science museum and once inside to pay selective attention to specific exhibitions or exhibit elements. Such opportunities to pursue topics of personal interest and the learning that results as a consequence can have important application to real-life situations. Museum professional Aubrey Tulley described an encounter with a young woman he met at a reception who ascribed great significance to her learning experience at the Science Museum in London (Tulley & Lucas, 1991). According to the woman, she had recently visited her sister who said that she was worried that her small children might leave the house unsupervised because the lock on the back door was broken. Upon hearing her sister's concern, the young woman borrowed a screwdriver and mended the lock. A week before this incident, she had spent time assembling the lock and key exhibit at the Science Museum. She asserted that without having encountered the exhibit she would never have had the confidence to respond to her sister's problem as she did. She described this as an emotional effect, on her "confidence," but clearly it was an emotional, cognitive, and psychomotor learning experience, a learning experience that she could transfer to a new situation.

It is significant that both this learning and the learning about bridges described earlier occurred not from a book or lecture, but at a museum. Both people learned while doing, while playing with locks, gears, and counterweights. And it is also significant that, at least in one case, the learning occurred while actively engaged in playing with a grandson. Successful science museum exhibitions, performances, films, programs, and Web sites share these common qualities, permitting the learner to seek the level of engagement and understanding appropriate for that person and tapping into personal and emotional entry points. A good exhibition, performance, or film can be understood at many different levels and from many different perspectives. Thus the learner is engaged in a variety of ways that make sense to them personally and can be challenged at a variety of different skill levels. Thus engagement, a flow experience, can result because there is sufficient depth to permit appropriate levels of challenge for a wide range of users.

SOME EVIDENCE

HIV/AIDS Travelling Exhibition

In the early 1990s, with support from the U.S. Centers for Disease Control, a consortium of prominent American science museums developed a travelling exhibition on HIV/AIDS, *What About AIDS*?[1] When the exhibition was being developed in the early 1990s, awareness and concern about the disease was very high. Although opinions differed widely on a number of HIV/AIDS-related issues, most Americans felt a strong need to learn more about the subject. Thus, although potentially controversial, at the time of its development it was clear that an exhibition on this topic was going to be widely perceived by the public as interesting and timely. Through front-end research (research designed to assess what people know about or are interested in knowing about a topic conducted at the beginning of the exhibition development process), it was also determined that most Americans possessed both a high degree of awareness about the HIV/AIDS epidemic and a reasonably high knowledge of basic "facts" related to HIV/AIDS. However, most Americans lacked detailed knowledge of the science underlying HIV/AIDS, for example, the nature of viruses and the workings of the human immune system, and most were totally unaware of the various permutations of prevention strategies.

A series of evaluation studies conducted by Dana Holland and John Falk revealed that, by all measures, this exhibition was a successful learning experience for most who visited it (Falk & Holland, 1993; Holland & Falk, 1994). Although not every visitor walked out of the exhibition knowing at least one, predetermined specific new fact or concept about HIV/AIDS, the exhibition afforded every visitor the opportunity to connect to the topic and learn something that was personally relevant to them; the data suggested that this learning was highly variable, emotion-laden, and extremely personal. The feedback loops described above were clearly being reinforced.

In-depth, open-ended interviews with people after they completed their visit to the *What About AIDS?* exhibition revealed how personally constructed and affectively influenced each visitors' learning was. Although the specifics of what a person learned was sometimes hard to predict, the relationship of that learning to the individual was predictable. More often than not, a visit to the exhibition significantly strengthened a visitor's prior understanding of the subject; only occasionally did they exit with a significantly new understanding. Very few of the visitors interviewed upon exiting *What About AIDS?* demonstrated evidence of a radically new view of the epidemic. However, what people learned was expressed in strong emotional ways that demonstrated the powerful relationship of what someone knew to what they felt about it:

> They [in the exhibit] were talking about coming in contact with people who have AIDS and they were saying that you are more . . . of a threat to them then they are to you. You know, because the virus, uh, something about how the viruses can't be treated with antibiotics, but bacteria can. I saw that on one of the things, and I said, that's very interesting because everyone is so scared of the person with AIDS, but the person with AIDS should be scared of you. (Female, teens)

> It made you aware, it made you really realize that it can happen to anybody, you know. And I think it also [explains], for people who don't know it, the three main [ways] you can get it. (Male, 20s)

These changed perceptions also persisted over time. Follow-up data collected 3 months after seeing the exhibition indicated that two thirds of the visitors claimed to have thought about the exhibition since their visit. Examples of what people said were

> I found myself still thinking about some of the things in the exhibit even weeks later. For example, I hate to say it, but the dice exhibit [probability of getting AIDS interactive] really made me think about who I go out with these days. (Female, 20s)

> Just the other day, I saw this piece on TV about AIDS and I was able to understand what they were talking about, the immune system in all, because of that exhibit. (Male, 30s)

Findings from the *What About AIDS?* summative evaluation not only demonstrated the strong relationship between learning and emotion, but also suggested that it was a highly successful learning experience for visitors with a wide variety of backgrounds, knowledge, and interest in the topic. Why? Researchers have recently argued that it would be difficult to design a more ideal setting for meaningful learning and meaning-making than a well-designed museum or science center (Dierking et al., 2002; Falk and Dierking, 2000). Increasing evidence suggests that when exhibitions and programs are designed well, they are very successful at supporting free-choice learning and public understanding of science and technology. I believe there are three reasons for this success:

1. Each visitor can experience the ideas/phenomena presented on their own terms, freely choosing what to attend to and interact with depending on their prior knowledge, interest, and experience

2. The ideas/phenomena presented are very accessible because for the most part these institutions are safe, comfortable, fun, social environments
3. The ideas/phenomena are embedded in rich real world contexts where visitors can see and directly experience the real world connections of these science and technology ideas/phenomena

As described earlier, this exhibition was developed in the early 1990s, when most Americans had not only heard about HIV/AIDS, but were also deeply concerned and interested in the topic. Although opinions differed widely on a number of HIV/AIDS-related issues, still many, if not most Americans, felt a strong need to know more about the subject, in other words they were motivated and interested in learning more. Using this information about potential visitors (a shared common interest, level of awareness, and range of understanding about the topic), it was possible to devise a strategy for how to present the topic of HIV/AIDS in an exhibition that would accommodate the interests and prior understanding of the vast majority of the public likely to encounter the exhibition. However, despite the rare confluence of *general* interest, awareness, and knowledge afforded by the subject of HIV/AIDS in the early 1990s, the *What About AIDS?* exhibition would not have been successful if it had been designed in a linear fashion, with a single entry and exit point. For though the public shared a general interest, awareness, and knowledge about HIV/AIDS, they did not share a *specific* interest, awareness, and knowledge about the topic. In other words, although two individuals may have been generally interested in HIV/AIDS and had roughly comparable awareness and knowledge, individual A, who was married, monogamous, and 67 years old might have been primarily interested in learning about how the epidemic might influence the health care system and the economy, while individual B, who was 19 years old and single, might have been primarily concerned with her chances of getting HIV/AIDS over the next few years.

A major design element of this exhibition was the incorporation of choice into the exhibition. Visitors could select between three general topics—biology of HIV/AIDS, HIV/AIDS as epidemic, and HIV/AIDS prevention—and then could also select from a multitude of specific topics: what is a virus, how does the immune system work, and what are the relative advantages and disadvantages of different birth-control methods on HIV/AIDS prevention. There were also choices for different learning modalities; one could read, watch video, manipulate hands-on interactives, use computer programs, and/or listen to audiotapes. A visitor also could choose between a variety of different approaches to the topic of HIV/AIDS, for example, they could learn about it through presentations of scientific facts and concepts, they could examine epidemiological charts and graphs, there were tapes and photographs detailing firsthand accounts of individuals with HIV/AIDS, and there were even opportunities for visitors to describe/share their own personal stories about HIV/AIDS. As a consequence, a wide range of visitors, each with their own diverse set of specific interests and knowledge, could select how and what they chose to learn about the topic.

The *What About AIDS?* exhibition also afforded very personal experiences. A family group could enter the exhibition, split up, and each member utilize a separate part of the exhibition, occasionally coming together to share notes and suggest parts of the exhibition for others to see or they could interact together throughout the exhibition. And the findings of the summative evaluation supported the effectiveness of this aspect of the exhibition. Not only did visitors find the content exceedingly accessible in a number of ways, with information presented from a variety of different perspectives, and at a variety of different points within the exhibition, observation data collected during the summative evaluation of the exhibition revealed that visitors had taken advantage of the choices offered. The developers appreciated that visitors entered and exited the exhibition with differing learning agendas and purposes, and strove to accommodate these differences.

The emotional component of this exhibition was also important to its success. As suggested earlier, emotion is a vital aspect of learning and problem solving and, consequently, it is an important dimension of many successful learning experiences. The topic of HIV/AIDS is a topic strongly infused with emotion and controversy, an aspect of the exhibition that caused some administrators and Boards of Directors around the United States great angst as the exhibition traveled. Although the emotional and controversial aspects of the exhibition may have been a political negative in some communities, these factors contributed to it being a very successful exhibition for personal learning. Being able to capitalize upon emotion is an important dimension of learning in free-choice settings such as museums and science centers. Fun, excitement, joy, mystery, sadness, surprise, pathos, anticipation, and empathy are all emotional experiences that can and should be considered fundamental constituents of learning. Education and enjoyment are not opposite ends of a continuum, they are separate and complementary, and in the museum context they combine to become the museum experience. Arguably, the essence of this experience is choice in what and when to learn; personal control over the learning.

FAMILY LEARNING INITIATIVE AT THE CHILDREN'S MUSEUM, INDIANAPOLIS

A systemic research effort at The Children's Museum (TCM) of Indianapolis, the Family Learning Initiative, provides two additional examples of how affect influences and in some cases *is* the learning. One study was designed to document the nature and extent of family interaction and engagement within a science exhibition about biological, medical, and cultural aspects of bones (Luke et al., 2002). Data collection included tracking family interactions, conducting interviews, utilizing Personal Meaning Mapping, a constructivist methodology developed and refined at the Institute, and recording conversations. Findings revealed that the exhibition succeeded in facilitating science learning about bones and enhancing both children's and adults' thinking about the biological, medical, and cultural aspects of bones. Data from the interviews and conversations revealed that the exhibition gave families opportunities to spend time together, relate exhibit content to their own personal family history, and build a collective identity. The data were replete with instances of family members excitedly sharing stories of broken bones (prompted by x-rays in the medical section of the exhibition) or discussing what the family dog might look like without its fur (prompted by skeletal models of animals in the zoo section of the exhibition). Families shared information and personal stories, working together to make meaning from the exhibition. These experiences also extended beyond the museum as many families went home to look at "Dad's x-rays" or to draw the dog without its fur.

In another Family Learning Initiative study, Luke and colleagues (2001) examined the long-term impact of two youth-based programs on young adults and their families, one focused on science exclusively, another a multidisciplinary program. Findings demonstrated that these programs influenced participants' attitudes, interests, and awareness. However, interestingly, these findings were not only observed at the individual level. Data gathered within a social systems approach demonstrated that these programs influenced family dynamics, giving young adults the opportunity to explore new roles, perspectives, and identities within the family and learn new things about family members. In some cases, interests that young adults developed within the program were carried over into the family context, influencing siblings and/or parents to pursue these same interests. Programs also influenced young adults' contributions and connections to the larger sociocultural community, fostering a tolerance of other people and cultures and cultivating a sense of civic responsibility, and helping youth to learn to participate successfully within a community of learners.

CONCLUSIONS

In this chapter I have suggested that free-choice learning arenas such as museums, science centers, zoos, aquariums, and the Internet are excellent environments, different windows so to speak, in which to explore relationships between affect and cognition. I've also tried to provide some examples of studies that demonstrate the potential of this approach and the notion that investigations in these settings might tease out and focus on the affective dimensions of learning in ways that are difficult in other settings. I hope that these findings will encourage researchers to take a more integrated approach to the investigation of emotion and cognition, not just in schools, but across all aspects of the science learning infrastructure (schools, universities, museums, science centers, etc.).

However, we still have much to learn about how people learn in free-choice settings such as museums, and how they connect these science-learning experiences to those they have at home and at school. Although the results shared in this chapter suggest that museums can foster the affective learning of the individual, as well as such learning at the family and community level, with the exception of the studies at the Children's Museum in Indianapolis, none of these studies were intentionally designed to investigate these issues specifically. What is needed at this stage are a series of research questions and ultimately studies, which would be set in "museums," build on the theoretical frameworks proposed here (the influence of emotion and other affective factors, motivation, and interest) and would more specifically target the influence of these factors on science learning in free-choice environments, as well as science learning in school and at home. Potential research questions might include

1. What role do emotions, motivation, interest, and other affective factors play in supporting free-choice learning in museums and other free-choice learning settings? Are these factors independent or interconnected?
2. Do factors such as emotions, motivation, and interest play equally important roles for all types of visitors/users or are their predictable patterns of influence for different "types" of visitors/users?

3. Do learners exhibit similar affective patterns within their learning across settings or do different settings result in different emotional and affective responses?
4. In what ways does the free-choice learning that occurs as a result of experiences in these settings relate to the science learning that results from experiences at home and in school?

Clearly, the promise that the earlier part of this chapter suggested has only been superficially investigated; however, there are additional signs that this is a fruitful area to pursue. In a recent dissertation study of family learning, in which families who were frequent visitors to science museums were observed in museums, at home, and in other leisure environments over the course of 18 months, the in-depth data revealed the ways in which families over time used seemingly unrelated interactions in museums and at home, to develop their family identity, in the case of these families, identities very emotionally centered on learning. Studying the families' interactions across multiple learning environments provided a needed lens for understanding the complex motivations underlying the families' practices in the museum (Ellenbogen, 2003). This is but one example of how this research could provide a new window with which to gaze through, a window that might elucidate even more clearly how the various learning experiences people engage in, support and facilitate their science learning across the different aspects and time span of their lives.

REFERENCES

Clifford, G. (1991). The past is prologue. In K. Cirincione-Coles (Ed.), *The future of education: Policy issues and challenges* (pp. 135–147). Beverly Hills, CA: Sage.

Covington, M. V. (1992). *Making the grade: A self-worth perspective on motivation and school reform.* Cambridge: Cambridge University Press.

Csikzentmihalyi, M. (1990a). *Flow: The psychology of optimal experience*. New York: Harper Collins.

Csikzentmihalyi, M. (1990b). Literacy and intrinsic motivation. *Daedalus, 119*(2), 115–140.

Csikzentmihalyi, M., & Hermanson, K. (1995). Intrinsic motivation in museums: Why does one want to learn? In J. Falk & L. Dierking (Eds.) *Public institutions for personal learning* (pp. 67–77). Washington, DC: American Association of Museums.

Csikzentmihalyi, M., & Nakamura, J. (1989). The dynamics of intrinsic motivation: A study of adolescents. *Research on motivation in education: Vol. 3. Goals and cognitions.* New York: Academic Press.

Damasio, A. R. (1994). *Descartes' error: Emotion, reasons, and the human brain.* New York: Avon Books.

deCharms, R. (1992). *Personal causation: The internal affective determinants of behavior.* New York: Academic Press.

Deci, E. L. (1992). The relation of interest to the motivation of behavior: A self-determination theory perspective. In K. A. Renninger, S. Hidi, & A. Krapp. (Eds.), *The role of interest in learning and development* (pp. 27–39). Hillsdale, NJ: Erlbaum.

Deci, E. L., & Ryan, R. M. (1985). *Intrinsic motivation and self-determination in human behavior.* New York: Plenum.

Deci, E. L., Schwartz, A. J., Sheinman, L., & Ryan, R. M. (1981). An instrument to assess adults' orientations toward control versus autonomy with children: Reflections on intrinsic motivation and perceived competence. *Journal of Educational Psychology, 73*, 642–650.

Deiner, C. I., & Dweck, C. S. (1980). An analysis of learned helplessness: The process of success. *Journal of Personality and Social Psychology, 31*, 674–685.

Dierking, L. D., Luke, J. J., & Büchner, K. S. (2002). Science & technology centers: Rich resources for free-choice learning in a knowledge-based society. *International Journal of Technology Management, 25*(5), 56–65.

Dierking, L. D., & Pollock, W. (1998). *Questioning assumptions: An introduction to front-end studies in museums*. Washington, DC: Association of Science-Technology Centers.
Ellenbogen, K. M. (2003). From dioramas to the dinner table: An ethnographic case study of the role of science museums in family life. Unpublished dissertation, Vanderbilt University, Nashville, TN.
Falk, J. H., & Dierking, L. D. (1992). *The museum experience*. Washington, DC: Whalesback Books.
Falk, J. H., & Dierking, L. D. (2000). *Learning from museums: Visitor experiences and the making of meaning*. Walnut Creek, CA: AltaMira.
Falk, J. H., & Holland, D. G. (1993). *What About AIDS Traveling Exhibition: Remedial evaluation*. Annapolis, MD: Institute for Learning Innovation. (formerly Science Learning)
Harlow, H. F. (1954). Motivational forces underlying behavior. In *Kentucky symposium, Learning Theory, and clinical research* (pp. 36–53). New York: Wiley.
Hidi, S. (1990). Interest and its contribution as a mental resource for learning. *Review of Educational Research, 60*, 549–571.
Holland, D. G., & Falk, J. H. (1994). *What About AIDS Traveling Exhibition: Summative evaluation*. Annapolis, MD: Institute for Learning Innovation. (formerly Science Learning)
James, W. (1950). *Principles of psychology* (2 vols.). New York: Dover. (Original work published 1890)
Luke, J., Cohen Jones, M., Wadman, M., Dierking, L. D., & Falk, J. H. (2002). *Phase II programs study: The Children's Museum of Indianapolis*. Technical report. Annapolis, MD: Institute for Learning Innovation.
Luke, J., Dierking, L., & Falk, J. (2001). *The Children's Museum of Indianapolis. Phase I report*. Technical report. Annapolis, MD: Institute for Learning Innovation.
Maehr, M. L. (1984). Meaning and motivation: Toward a theory of personal investment. In *Research on motivation in education: Vol. 1. Student motivation* (pp. 216–232). New York: Academic Press.
McCombs, B. L. (1991). Motivation and lifelong learning. *Educational Psychologist. 26*(2), 117–127.
Paris, S. G. (1997). Situated motivation and informal learning. *Journal of Museum Education. 22*(2/3):22–27.
Paris, S. G., & Cross, D. R. (1983). Ordinary learning: Pragmatic connections among children's beliefs, motives, and actions. In J. Bisanz, G. Bisanz, & R. Kail (Eds.), *Learning in children* (pp. 137–169). New York: Springer-Verlag.
Piaget, J. (Trans. and Ed. by T. A. Brown and C. E. Kaegi). (1981). Intelligence and affectivity. Their relationship during child development. *Annual reviews monograph*. Palo Alto, DA: Annual Reviews.
Pintrich, P., & DeGroot, E. (1990). Motivational and self-regulated learning components of classroom academic performance. *Journal of Educational Psychology. 82*, 33–40.
Ramey-Gassert, L., Walberg, H. J. III, & Walberg, H. J. (1994). Reexamining connections: Museums as science learning environments. *Science Education, 78*(4), 345–363.
Rohrkemper, M., & Corno, L. (1988). Success and failure on classroom tasks: Adaptive learning and classroom teaching. *Elementary School Journal, 88*, 297–312.
Rose, S. (1993). *The making of memory: From molecules to mind*. New York: Anchor Books/Doubleday.
Rosenfield, I. (1990). *The invention of memory*. New York: Basic Books.
Schiefele, U. (1991). Interest, learning and motivation. *Educational Psychologist, 26*(3/4), 299–323.
Sylwester, R. (1995). *In celebration of neurons*. Alexandria, VA: Association for Supervision and Curriculum Development.
Tulley, A., & Lucas, A. M. (1991). Interacting with a science museum exhibit: Vicarious and direct experience and subsequent understanding. *International Journal of Science Education, 13*, 533–542.
White, R. W. (1959). Motivation reconsidered: The concept of competence. *Psychological Review, 66*, 297–333.

NOTE

[1] Members of the National AIDS Exhibit Consortium were: National Museum of Health & Medicine, The Franklin Institute Science Museum, Maryland Science Center, Exploratorium, Museum of Science & Industry (Chicago), California Museum of Science and Industry and Museum of Science (Boston) (cf., Aprison, 1993).

CHAPTER 10

MICHALINOS ZEMBYLAS

EMOTIONS AND SCIENCE TEACHING: PRESENT RESEARCH AND FUTURE AGENDAS

INTRODUCTION

Many educators and researchers point out that affective issues are important in teaching; however, little has been done to incorporate affective concerns in a systematic way in research on teaching. As Norman (1981) pointed out two decades ago, most cognitive theorists preferred to ignore the affective domain and concentrate instead on developing information-processing models of purely cognitive systems. Such an approach has been most obvious in the investigation of teacher cognition and teacher beliefs (e.g., Clark & Peterson, 1986; Kagan, 1992a, 1992b; Nespor, 1987; Richardson, 1996). The emphasis on teacher beliefs has been on teachers' views and perspectives often without any discussion of the relevance of those beliefs to teacher emotions. For the most part, this area of research still avoids addressing how teacher beliefs interact with teacher emotions and attitudes or what the role of teacher emotions is in understanding teaching and learning.

Yet, conducting research on emotions in education or on the emotions of teachers, more specifically, presents several challenges. First, emotions are very fluid and much more complex and difficult to describe than cognition (Boler, 1999; Janack, 2000; McLeod, 1989; Simon, 1982; Zembylas, 2002a, 2002b). Second, one of the reasons for the neglect of investigating emotion in teaching may be due in part to the domination of cognitive psychology over educational research, and the difficulty in capturing the emotional components of teaching for research purposes. Finally, there is the legacy of dualism, which has opposed reason to emotion, and accorded reason the high status inscribed in Western thinking. Embedded in Western thought is the assumption that emotions threaten the disembodied, detached, and neutral knower; consequently, as it is suggested, emotions do not offer any valid knowledge. This view has placed emotions in an inferior role and made much more difficult the legitimation of research on teacher emotion. This uneasy relation between emotion and reason provides the social and historical context from which many current views on emotions continue to emerge (Schutz & DeCuir, 2002). These issues assert that we are convinced about the importance of emotions in education only at a general abstract level (Beck & Kosnik, 1995).

One can imagine, then, what happens in science education when educators in general have a hard time accepting emotions as a legitimate subject of research. After all, in schools, science is portrayed as rational and non-emotional (Alsop, 2001; Zembylas, 2002a). What is the place of emotion in teaching and learning science? Why should one pay attention to the way teachers feel about science and science teaching?

Steve Alsop (ed.) Beyond Cartesian Dualism, 123–132.

The chapters in this book suggest an increased interest in research on emotions in science education. This chapter takes up the case for teacher emotions in science education. My argument is that if science educators aspire to a rich understanding of science teaching, they can ill afford to leave out the emotional aspects. As Mandler (1989) argued more than a decade ago: "[I]t is not enough for cognitive models to have nodes or processors labeled *fear* or *joy* that can be accessed whenever appropriate. That approach simply acknowledges theoretical impoverishment; it does not solve the problem of how emotions arise or how they are to be represented" (p. 4, original emphasis). Recent theoretical and empirical work in the social sciences has demonstrated the symbiotic nature of the relationship between cognition and emotion (Planalp & Fitness, 1999). It would be unwise of us, science educators, to ignore the role of teacher emotion in science teaching.

The purpose of this chapter is to evaluate current research on the role of emotions in teaching in general, and in science teaching, more specifically, and propose some possible directions for future investigations of teacher emotion in science education. For this purpose, I draw heavily on my own empirical studies on teacher emotion in science teaching. First, I provide a brief summary of teacher emotions and their role in (science) teaching. Then I suggest some future theoretical and methodological possibilities for research in this area and close with a call for a more integrated understanding of emotion and cognition in science education.

EMOTIONS AND (SCIENCE) TEACHING: PRESENT RESEARCH

A review of past attempts by educators to explore the importance of teacher emotion highlights the "*neglect* of a topic which is of daily concern to practitioners... [because] as an occupation teaching is highly charged with feelings, aroused by and directed towards not just people but also values and ideas" (Nias, 1996, p. 293, added emphasis). Nias identifies how problematic this lack of research is about teacher emotion:

> Despite the passion with which teachers have always talked about their jobs, there is relatively little recent research into the part played by or the significance of affectivity in teachers' lives, careers and classroom behaviour. Since the 1960s teachers' feelings have received scant attention in professional writing. At present, they are seldom systematically considered in pre- or in-service education. By implication and omission teachers' emotions are not a topic deemed worthy of serious academic or professional consideration. (1996, p. 293)

During the last two decades, however, there have been an increasing number of studies illustrating the role of teacher emotion in curriculum and teaching. Elsewhere, I (Zembylas, 2003a) describe two waves of research on teacher emotion. As researchers in education began to recognize the power of emotion in teaching and asked what schools could do to take advantage of it, a first wave of research (roughly the period in the 1980s and early 1990s) focused on establishing awareness of the role of emotions in teaching (e.g., see Nias, 1989; Salzberger-Wittenberg et al., 1983). The second wave of research on teacher emotion covers the period during the last decade and focuses on the idea of social relationships, recognizing emotion as part of relationships in the classroom and the school context (e.g., see Blackmore, 1996; Golby, 1996; Hargreaves, 1998a, 1998b, 2000a, 2000b; Jeffrey & Woods, 1996; Kelchtermans, 1996; Lasky, 2000; Little, 1996; Schmidt, 2000; Tickle, 1996).

The central argument of the relatively few studies undertaken during the first wave of research on teacher emotion was that "effective teaching and learning is necessarily affective, that it involves human interaction, and that the quality of teacher–pupil relationships is vitally important to the learning process" (Osborn, 1996, p. 455). This first wave of research on teacher emotion focused also on the broad ideas of stress and burnout, although the term "emotion" was almost never used to theorize teachers' experiences of career (Zembylas, 2003a). With the exception of discussions about "fatigue," "frustration," and "nervous tension," researchers and theorists did not give much attention to how teacher emotion was part of the school culture (e.g., see Dworkin, 1987; Farber, 1991; Truch, 1980; for more recent work see Cherniss, 1995; Vandenberghe & Huberman, 1999).

The authors of studies in this first wave provide an indication of the importance of considering emotions in teaching and learning (Zembylas, 2003a). These studies hint at a problem that will surface and be discussed in research conducted during the second wave. It involves the interaction of teacher emotion with other dimensions in teaching such as teacher performance, teacher knowledge, and the social and political context of the classroom and the school. The recommendations from this work point to a need for teachers to examine their emotions and to negotiate an effective teaching role based on emotional experiences gained with the students. The research on the impact of teacher emotion regarding student learning appeared to be rather preliminary (Zembylas, 2003a).

Theorists and researchers of the second wave, inspired primarily by sociological thought (in particular, social constructionism), suggest that educators construct an array of positive and negative emotions in their teaching practice and assert that the centerpiece of this research is an exploration of the social interactions among teachers, students, parents, and administrators. Emotions at the individual level are increasingly recognized as governed by social interactions. A teacher's emotions, then, are determined not only or even primarily by internal individual (intrapersonal) characteristics, but rather by relationships, a view known as the social constructionism of emotion (e.g., see Harré, 1986). As part of this social construction trend in research on teacher emotion, feeling and emotional display in teaching can be argued not only to manifest but also to reify relationships in the classroom and the school context (Zembylas, 2001, 2002b, 2003a, 2003b).

A major feature of the studies during the second wave is that they document the complexity and range of emotions held by teachers. The area appears fragmented with a wide array of avenues being pursued—the area is new, thus one expects a certain degree of fragmentation—but mostly those avenues are limited within sociological and psychological frameworks (two of the strongest albeit not the only traditions in studying emotions). This corpus of research during the second wave, despite its weaknesses, has advanced the study of emotion in teaching in important ways (Zembylas, 2003a). Let us consider in more details some of these studies undertaken in the area of science education.

TEACHER EMOTION IN SCIENCE EDUCATION

It is encouraging that in the last few years science educators have investigated the relationship between teaching and emotion (e.g., Barker 2001; Zembylas 2001, 2002a, 2004a, 2004b; Zembylas & Barker, 2002) and why this is important to science teachers (e.g., Alsop, 2001; Matthews, 2002; Matthews et al., 2002; Watts & Alsop, 1997; Watts & Walsh, 1997). My 3-year ethnographic case study (Zembylas, 2001, 2002a), of an experienced early childhood teacher, details the ways in which teacher emotion can contribute both to the educational experiences of children and to the professional experiences of the teacher herself. I argue that positive and negative emotions play a significant role in a teacher's construction of her science pedagogy, curriculum planning, and relationships with children and colleagues.

This work emphasizes how the emotional aspects of the science teacher-self in becoming or being a science teacher, the acquisition and use of pedagogical approaches, and the application of professional judgment in practice are inextricably linked. I suggest, "If we want progress in science education, we need to look more carefully at the emotions of science teaching, both negative and positive emotions, and use this knowledge to improve the working environment of science teachers," and that "when the emotional aspects of science teaching and science teacher development are considered seriously, it is safe to say that what is at stake in science teacher education and science curriculum reform and how best to enrich them, will never look the same again" (Zembylas, 2002a, p. 98). My work develops a conceptual and methodological framework that is based on an interdisciplinary approach in researching emotions. Theoretically, although, it builds on social constructionism and brings alternative theoretical tools such as Foucaultian genealogy.

For this reason, I define my work under the heading: genealogies of emotions in science teaching. More specifically, my aim is to explore the conditions under which teachers' emotions in science teaching are shaped and performed, to discover how they might be "disciplined," to destabilize and denaturalize the regime that demands the expression of certain emotions and the disciplining of others, and to elucidate the "emotional rules" that are imposed and the boundaries entailed by those rules (Zembylas, 2003b). The place of emotion in science teacher self-formation plays a central role in the circuits of power that constitute some teacher-selves while denying others. Critically understanding these processes of discipline and domination, I suggest, is crucial if we are to promote the possibility of new forms of subjectivity in science education.

Similarly, my recent co-authored study of preservice teacher emotions (Zembylas & Barker, 2002) examines the power of analytical tools such as *individual spaces* and *community conversations* in creating spaces of *emotional comfort* that support the efforts of preservice teachers to become reflective practitioners in their teaching of science. This study highlights the significance of creating emotionally supportive environments for the development of positive attitudes and professional knowledge of preservice teachers in science education. By understanding preservice teachers' attitudes and emotions and the relationship of these to their understanding of science and science teaching, we point out, that science educators will be in a better position to design more effective programs and create supportive environments to recruit and retain more teachers.

Other studies by Alsop (2001), Watts and Alsop (1997), and Watts and Walsh (1997) deal with the emotional aspects of science learning and the consequences for the teacher. Although their focus is not the science teacher *per se*, these studies raise a number of important questions that are relevant to the need for teachers to have greater awareness of the emotional components of science teaching and learning.

In general, what seems clear is that the relationship between science teaching and emotion deserves further attention. The few studies conducted so far emphasize that science educators and science teachers should consider the effects of the emotional aspects of learning and teaching science in planning effective instruction. In fact, according to these researchers, attention to the role of emotion in science teaching and learning may be needed in order to accomplish the goal of positive attitudes toward science, a goal viewed by science educators as an important outcome of science teaching (e.g., American Association for the Advancement of Science [AAAS], 1993; National Research Council [NRC], 1996).

Many teachers—preservice and in-service—feel uncomfortable dealing with some of the emotional aspects of teaching science (Nichols et al., 1997; Richmond et al., 1998). Even teachers who are comfortable discussing their emotions of discomfort and anxiety often feel inadequately prepared to deal with how they feel (Zembylas, 2002a, 2004a). Perhaps these reactions are appropriate to the current situation given the complex relationship between emotion and teaching, on the one hand, and the contemporary pressures, declining job satisfaction and occupational stress in teaching (e.g., Farber, 1991; Troman & Woods, 2000; Vandenberghe & Huberman, 1999), on the other hand. These are certainly issues that need to receive more attention in investigations of teacher emotions in science education.

FUTURE RESEARCH AGENDAS ON TEACHER EMOTION IN SCIENCE TEACHING

Although the research on teacher emotion in science education has made some progress toward understanding the role of emotion in teaching and learning science, much work is needed. One thing that is clear so far is that science teaching practice is necessarily affective and involves an incredible amount of emotional labor. The emotional dissonance created by emotional labor can arguably lead to stress and burnout and there is now a considerable body of work that links teacher stress with teachers' early exit from the profession (e.g., see Huberman, 1993; Travers & Cooper, 1996). Given this realization, the study of teacher emotion becomes an important area of research in science curriculum and teaching.

Future studies can focus on different ideas such as the influence of teacher emotion on one's self-concept, perception, and judgment, the relationship between emotion and teacher identity, how students are influenced by teachers' emotions, and how teacher emotions influence curricular decisions and curriculum reform, all in the context of science education. Apart from the notion that teacher emotion affects curricular decisions and teacher knowledge, one can argue that a number of fundamental problems remain unresolved in the area of research on teacher emotion not only in science education but also in education, more generally. These problems are constitutive aspects of possible directions for research.

First of all, the theorization of teacher emotion, especially in the second wave, is mostly inspired either by a sociological (interpersonal) framework—especially social constructionist and contextualized perspective—or a psychological framework. Missing is an exploration of teacher emotion as embedded in school culture, ideology, discourse, and power relations. Very few studies have paid attention to political and cultural issues—e.g., how different practices establish and regulate emotional rules and require emotion management in the context of curriculum and teaching. My previous work with Barker (Barker, 2001; Zembylas, 2001, 2002a, 2004a; Zembylas & Barker, 2002) appears to be an exception but there needs to be further evidence of these claims since this work is based on an examination of a few case studies.

For instance, it may be useful to examine the discourses used to discuss emotions in science teaching and learning from a poststructuralist perspective. Researchers in science education so far have put considerable emphasis on either psychological or sociological approaches in their studies. Would it not be useful to employ different research approaches in exploring emotions in science education? Another problem that needs to be explored within a sociocultural framework is the impact of testing on students and teachers. Because standardized testing—especially in science and mathematics—has become a central focus of many science curricula in the United States, it is important to understand how such an emphasis influences science teaching and learning emotionally.

Further, recent studies on teacher emotion are mostly drawing their theoretical framework from social constructionist perspectives and explore teacher emotion as embedded in social interactions (e.g., see Hargreaves' work). However, this reliance on social constructionist ideas to theorize about teacher emotion is somewhat problematic. For one thing emotions have also crucial affective components (Stocker, 1996) and cannot be understood merely in terms of beliefs or desires and how they are socially constructed. This becomes more obvious in cases of radical personal transformations (e.g., "conversion experiences"). These transformations are not simply changes in beliefs, because emotions are embodied experiences and it is this affective part of emotions that is neglected in attempts to define emotions simply in terms of beliefs and judgments (Janack, 2000). Explaining teacher emotion in rationalist terms that reduce social interactions to being driven either by desire for rationality or by need for a coherent belief system has resulted in ignoring the role of affectivity and emotionality. A theorization of emotion as embodied empowers attempts to study its social, cultural, and political aspects.

In addition, many researchers discuss the role of emotion in cultural learning from an anthropological point of view, and suggest that both beliefs and emotions are useful in cognition (D' Andrade, 1981; Reddy, 1999; Rosaldo, 1984). Research in various classroom emotional cultures alerts educators to the limitations of our lay and professional vocabularies for fully integrating emotion and cognition together in the ways we think about education. Both cognition and emotion are characteristics of human beings. The best that educators can do is to understand how these terms come about in science teaching, how they interact, and how their inevitable symbiosis can be put to the use of our teachers.

Thus, in general, there is clearly a need to study teacher emotion in the context of classrooms and schools, taking historical, political, cultural, and social influences into consideration. This approach is necessary because to understand the role of teacher emotions in science education, it is important that we understand the context in which these emotions are enacted. Such an approach may include the construction of more "genealogies of emotions in science teaching" (Zembylas, 2002a).

Also, a research question that has promise is how emotions affect successes and failures in science teaching? Most importantly, a theorization of teacher emotion in science teaching can be developed to illustrate the role of emotion in establishing and maintaining a teacher's self-esteem. This is particularly significant given the existing documentation that elementary teachers often lack confidence in their science knowledge and skills and thus avoid teaching science (e.g., Nichols & Tippins, 2000; Zembylas, 2004a; Zembylas & Barker, 2002).

Another issue for consideration in future research in this area is that it is not yet clear what research methodologies should be used to explore teacher emotion. For one thing, there is quite obviously a difference between the expression of emotion and the emotion itself. Traditional questions that served as the site of fierce debates in other disciplines were how does one legitimately move from what is observed (the expression) to what cannot be observed (the emotion)? How does one know, on the basis of the varied expression, that the difference is one of emotion rather than merely expression? (Solomon, 1984). These questions have not been adequately theorized in the area of research on teacher emotion not only in science education but more generally as well. Also, there is a need for using multiple methods in the study of teacher emotion in science education such as, a combination of quantitative and qualitative methods or the use of longitudinal studies.

Finally, another issue that remains unresolved in the area of research on teacher emotion is the need to develop pedagogies that promote empowerment and teacher self-development in science education. Most work so far has been descriptive in terms of identifying factors relevant to teacher emotion and how these influence curriculum and (science) teaching. Central to developing such pedagogies are ideas that account for the intersections of emotions, power, and ideology. On the basis of this notion, teachers' emotions cannot be regarded only in their interpersonal aspects; instead they need to be regarded as the very location of the capacity to embrace, revise, or reject discursive practices of whatever kind.

For example, theoretical innovations by feminist, poststructuralist, and postcolonial writers offer important directions away from a depoliticized and dehistoricized analysis of emotion. Unless one specifies the ground—by saying that emotions are embedded in power relations and ideology—one cannot have a politics of emotion. Acknowledging that teacher emotion is the very site of the capacity to affect change, as past research has shown us, there is significant potential in feminist and poststructuralist ideas of emotion to provide new tools for theorizing about teacher emotion and for using this theorization to initiate and to sustain teacher self-transformation in science education. Such theorization has the potential of empowering teachers to subvert emotions of vulnerability and powerlessness they experience in science teaching.

A LAST WORD

The criticisms I have offered for past attempts in the area of research on teacher emotion are not meant to undermine the value of this work. Those who have argued for the importance of emotions in teaching and how emotions are social constructions within relationships have made an important contribution both by posing the issues in sharp terms and by offering rich and insightful accounts inspired by a coherent argument. But I suggest that new ideas are needed in enriching theorization about teacher emotion generally and in science education specifically and in conceptualizing the political stakes in clear and powerful ways. One should not forget that the main goal of research in this area is the improvement of science instruction and student learning. Also, a more integrated understanding of emotion and cognition in science education would be beneficial for how we think about teaching and learning science. Conversations about where we presently are in our research, as well as problematic areas of research on teacher emotion in science education would vastly broaden the development of theories and pedagogies that recognize teacher emotion as a site of empowerment and teacher self-development.

REFERENCES

Alsop, S. (2001). Seeking emotional involvement in science education: Food-chains and webs. *School Science Review, 83*(302), 63–68.

American Association for the Advancement of Science. (AAAS). (1993). *Benchmarks for science literacy*. New York: Oxford University Press.

Barker, H. B. (2001). A room of one's own: Concrete and conceptual spaces. In M. Osborne & A. Barton (Eds.), *Teaching science in diverse settings: Marginalized discourses and classroom practice* (pp. 59–78). New York: Peter Lang.

Beck, C., & Kosnik, C. M. (1995). Caring for the emotions: Toward a more balanced schooling. In A. Neiman (Ed.), *Philosophy of education* (pp. 161–169). Urbana, IL: Philosophy of Education Society.

Blackmore, J. (1996). Doing 'emotional labour' in the education market place: Stories from the field of women in management. *Discourse: Studies in the Cultural Politics of Education,17*, 337–349.

Boler, M. (1999). *Feeling power: Emotions and education*. New York: Routledge.

Cherniss, C. (1995). *Beyond burnout: Helping teachers, nurses, therapists and lawyers recover from stress and disillusionment*. New York: Routledge.

Clark, C., & Peterson, P. (1986). Teachers' thought processes. In M. Wittrock (Ed.), *Handbook of research on teaching* (3rd ed., pp. 255–296). New York: Macmillan.

D' Andrade, R. G. (1981). The cultural part of cognition. *Cognitive Science, 5*, 179–195.

Dworkin, A. G. (1987). *Teacher burnout in the public schools: Structural causes and consequences for children*. Albany, NY: State University of New York Press.

Farber, B. A. (1991). *Crisis in education: Stress and burnout in the American teacher*. San Francisco: Jossey-Bass.

Golby, M. (1996). Teachers' emotions: An illustrated discussion. *Cambridge Journal of Education, 26*, 423–434.

Hargreaves, A. (1998a). The emotional practice of teaching. *Teaching and Teacher Education, 14*, 835–854.

Hargreaves, A (1998b). The emotional politics of teaching and teacher development: With implications for educational leadership. *International Journal of Leadership in Education, 1*, 315–336.

Hargreaves, A. (2000a). *Emotional geographies of teaching and educational change*. Paper presented at the annual meeting of the American Educational Research Association, New Orleans, LA.

Hargreaves, A. (2000b). Mixed emotions: Teachers' perceptions of their interactions with students. *Teaching and Teacher Education, 16*, 811–826.

Harré, R. (Ed.). (1986). *The social construction of emotions*. New York: Basil Blackwell.

Huberman, M. (1993). *The lives of teachers*. London: Cassell.

Janack, M. (2000). Emotion and conversion. *The APA Newsletters, 99*, 184–186.

Jeffrey, B., & Woods, P. (1996). Feeling deprofessionalized: The social construction of emotions during an OFSTED inspection. *Cambridge Journal of Education, 26*, 325–343.

Kagan, D. (1992a). Implications of research on teacher belief. *Educational Psychologist, 27*, 65–90.

Kagan, D. (1992b). Professional growth among preservice and beginning teachers. *Review of Educational Research, 62*, 129–170.

Kelchtermans, G. (1996). Teacher vulnerability: Understanding its moral and political roots. *Cambridge Journal of Education, 26*, 307–324.

Lasky, S. (2000). The cultural and emotional politics of teacher-parent interactions. *Teaching and Teacher Education, 16*, 843–860.

Little, J. W. (1996). The emotional contours and career trajectories of (disappointed) reform enthusiasts. *Cambridge Journal of Education, 26*, 345–359.

Mandler, G. (1989). Affect and learning: Causes and consequences of emotional interactions. In D. McLeod & V. Adams (Eds.), *Affect and mathematical problem solving: A new perspective* (pp. 3–19). New York: Springer-Verlag.

Matthews, B. (2002). Why is emotional literacy important to science teachers? *The School Science Review, 84*(305), 97–104.

Matthews, B., Kilbey, T., Doneghan, C., & Harrison, S. (2002). Improving attitudes to science and citizenship through developing emotional literacy. *The School Science Review, 84*(306), 103–114.

McLeod, D. (1989). The role of affect in mathematical problem solving. In D. McLeod & V. Adams (Eds.), *Affect and mathematical problem solving: A new perspective* (pp. 20–36). New York: Springer-Verlag.

National Research Council. (1996). *National science education standards.* Washington, DC: National Academy Press.

Nespor, J. (1987). The role of beliefs in the practice of teaching. *Journal of Curriculum Studies, 19,* 317–328.

Nias, J. (1989). *Primary teachers talking: A study of teaching and work.* London: Routledge.

Nias, J. (1996). Thinking about feeling: The emotions in teaching. *Cambridge Journal of Education, 26*, 293–306.

Nichols, S., & Tippins, D. (2000). Prospective elementary science teachers and biomythographies: An exploratory approach to autobiographical research. *Research in Science Education, 30*, 141–153.

Nichols, S. E., Wieseman, K., & Tippins, D. (1997). A toolkit for developing critically reflective science teachers. *Research in Science Education, 27*, 175–194.

Norman, D. A. (1981). Twelve issues for cognitive science. In D. A. Norman (Ed.), *Perspectives on cognitive science* (pp. 265–295). Norwood, NJ: Ablex.

Osborn, M. (1996). Book reviews: The highs and lows of teaching: 60 years of research revisited. *Cambridge Journal of Education, 26*, 455–461.

Planalp, S., & Fitness, J. (1999). Thinking/feeling about social and personal relationships. *Journal of Social and Personal Relationships, 16*, 731–750.

Reddy, W. M. (1999). Emotional liberty: Politics and history in the anthropology of emotions. *Cultural Anthropology, 14*, 256–288.

Richardson, V. (1996). The role of attitudes and beliefs in learning to teach. In J. Sikula (Ed.), *The handbook of research in teacher education* (2nd ed., pp. 102–119). New York: Macmillan.

Richmond, G., Howes, E., Kurth, L., & Hazelwood, C. (1998). Connections and critique: Feminist pedagogy and science teacher education. *Journal of Research in Science Teaching, 35*, 897–918.

Rosaldo, M. (1984). Toward an anthropology of self and feeling. In R. Shweder & R. Levine (Eds.), *Culture theory: Essays on mind, self, and emotion* (pp. 137–157). New York: Cambridge University Press.

Salzberger-Wittenberg, I., Henry, G., & Osborne, E. (1983). *The emotional experience of teaching and learning*. London: Routledge & Kegan Paul.

Schmidt, M. (2000). Role theory, emotions, and identity in the department headship of secondary schooling. *Teaching and Teacher Education, 16*, 827–842.

Schutz, P., & DeCuir, J. T. (2002). Inquiry on emotions in education. *Educational Psychologist, 37*, 125–134.

Simon, H. A. (1982). Comments. In M. S. Clark & S. T. Fiske (Eds.), *Affect and cognition, The Seventeenth Annual Carnegie Symposium on Cognition* (pp. 333–342). Hillsdale, NJ: Lawrence Erlbaum.

Solomon, R. C. (1984). Getting angry: The Jamesian theory of emotion in anthropology. In R. A. Shweder & R. A. Levine (Eds.), *Culture theory: Essays on mind, self, and emotion* (pp. 238–254). Cambridge: Cambridge University Press.

Stocker, M. (1996). *Valuing emotions.* Cambridge, MA: Cambridge University Press.

Tickle, L. (1996). New teachers and the emotions of learning teaching. *Cambridge Journal of Education, 21*, 319–329.

Travers, C. J., & Cooper, C. L. (1996). *Teachers under pressure: Stress in the teaching profession*. London: Routledge.
Troman, G., & Woods, P. (2000). Careers under stress: Teacher adaptations at a time of intensive reform. *Journal of Educational Change, 1*, 253–275.
Truch, S. (1980). *Teacher burnout and what to do about it.* Novato, CA: Academic Therapy.
Vandenberghe, R., & Huberman, A. M. (1999). *Understanding and preventing teacher burnout: A sourcebook of international research and practice*. Cambridge: Cambridge University Press.
Watts, M., & Alsop, S. (1997). A feeling for learning: Modeling affective learning in school science. *The Curriculum Journal, 8*, 351–365.
Watts, M., & Walsh, A. (1997). Affecting primary science: A case from the early years. *Early Childhood and Care, 129*, 51–61.
Zembylas, M. (2001). A paralogical affirmation of emotion's discourse in science teaching. In M. Osborne & A. Barton (Eds.), *Teaching science in diverse settings: Marginalized discourses and classroom practice* (pp. 99–128). New York: Peter Lang.
Zembylas, M. (2002a). Constructing genealogies of teachers' emotions in science teaching. *Journal of Research in Science Teaching, 39*, 79–103.
Zembylas, M. (2002b). "Structures of feeling" in curriculum and teaching: Theorizing the emotional rules. *Educational Theory, 52*, 187–208.
Zembylas, M. (2003a). Caring for teacher emotion: Reflections on teacher self-development. *Studies in Philosophy and Education, 22*, 103–125.
Zembylas, M. (2003b). Interrogating "teacher identity": Emotion, resistance, and self-formation. *Educational Theory, 53*, 107–127.
Zembylas, M. (2004a). Emotion metaphors and emotional labor in science teaching. *Science Education, 55*, 301–324.
Zembylas, M. (2004b). Young children's emotional practices while engaged in long-term science investigations. *Journal of Research in Science Teaching, 41*(7), 693–719.
Zembylas, M., & Barker, H. (2002). Beyond "methods" and prescriptions: Community conversations and individual spaces in elementary science education courses. *Research in Science Education, 32*, 329–351.

SECTION THREE: PEDAGOGICAL INTERVENTIONS

OVERVIEW

In Section 3 our discussions have an applied nature; they describe and analyze particular approaches that have sought to explicitly incorporate emotions in practice. Here, the discussions focus on better understanding the affective dimensions of educational approaches situated with specific learning contexts.

The diversity of instructional approaches and contrasting pedagogical settings, by any measure, is impressive. Early on we are introduced to both dilemma and contradiction. The desire to cover the curriculum and ensure examination success seems to be, in many ways, at odds with creative practices that seek to nurture the wonder, joy, passion, and beauty in learning. The beauty of learning and the compassion of schooling contrast with the brutality of isolation and the loneliness of beginning to teach.

Discussion of the desire to reinvigorate and to reinvent science by emotionally challenging learners juxtaposes with the deep-routed need for stability and rebuilding. In science education, the debate is often about challenging images of science as certainty and authority. And yet, in the context of extreme emotional upheaval, it appears that these very elements might act as a platform to rebuild a life in turmoil. The notion that the world might somehow be predictable through science, serves—in Chapter 11—as an epistemological and ontological staring point to restore psychosocial resilience. In what follows there is discussion of the joys of discovery and transformation as well as the despair of social upheaval, failure, and underachievement.

The four case studies also offer empirical diversity; the teacher–researchers grapple with the ephemeral, transitory nature of affect and the tensions associated with making the internal visible and the "irrational" rational. Of course, there is a widely held assumption that while cognition is eminently quantifiable, affect is beyond description and comment with any meaningful specificity. While the effects of emotions can be observed and recorded externally, physiologically (increased heart beat, body temperature, and so on), and expressively (changes in facial expression), the inner emotion, of course, will always remain private. There is no "window into the mind," as Piaget once famously remarked. In the following four cases, the authors deploy both quantitative and qualitative methods to make sense of pedagogy through an interpretation of outward expression and action.

We start with child refugees. Frederic Perrier, in Chapter 11, describes a pedagogical interventional in a setting that few of us can (and probably will never) comprehend: refugee camps following genocide. In this context, active investigative science offers a unique and efficacious platform to rebuild a connection with a lost world. Perrier's chapter is an extended reflection and evaluation from the field, an exploration of the pragmatics and moral dilemmas associated with being an educator in a distant and

Steve Alsop (ed.) Beyond Cartesian Dualism, 135–136.

troubled land. In his personalised and emotional account, science education surfaces as a way of holding pattern and restoring predictability: a therapeutic means of re-establishing developmental patterns in learning after extreme trauma. The therapeutic properties of science education offer new research directions. Might the detached, anodyne, and positivistic qualities of science education (features that have been widely criticised throughout this text) be serving to support learning? Might the introduction of uncertainly in science education serve to disenfranchise some learners? Does science offer a secure way of connecting with the outside world?

In search of consilience, a way of combining science and arts, Mike Watts turns to expressive creative writing. Although normally the bedrock of the humanities, his claim is that poetry has much to offer science education as a means of releasing imagination and mystery through the "orchestration of confluence" (the integration of intellect, emotion, and behaviour in the teaching and learning of science). His eloquent discussion, serves to fuse Apollonian (critical–rationalist) and Dionysian (creative–passionate) desires. The basis of reinventing and reinvigorating science education Watts suggests is to be found in a better confluence of intellectual, emotional, and practical domains. Many, I suggest, would agree.

Bonnie Shapiro, in Chapter 13, offers a compelling account of teacher transformation: A mentor's personalised reflection of the dynamic interplay between meaning and identity. The role of practicum supervisor offers a vantage point to reflect on a student teacher's passage from despair to success. Here, challenges inherent in learning to teach serve to galvanise personal sources of insight and scaffold teacher prosperity. Shapiro's compassionate account underscores the autonomous nature of teaching transformation grounded in a sense of personal identity and investment. There is much to draw from her account, as I discuss in my opening chapter.

The final chapter is left to Brian Matthews, one of my previous professors, to explore conceptual, social, and emotional growth. Here, we turn to engaging affect as a way of addressing and challenging inequality. The ISED (Improving Science and Emotional Development) project, based at Goldsmith's College London, aims to foster informed pedagogy with emotional as well as social developmental goals (see Matthews, 2004). Data drawn from a small-scale quasi-scientific comparative study is used to evaluate the pedagogical approaches advocated. While, the individualistic and decontextualised nature of emotional intelligence is not without controversy (see for example Boler, 1999), Matthew's work serves nicely to raise debate about the moral and ethical dilemmas associated with emotional development. The final chapter, for me, offers a fascinating classroom intervention that raises futuristic concerns of quantification and accountability. What emotions count? Whose emotions count? What counts as emotional development?

REFERENCES

Boler, M. (1999). *Feeling power: Emotions and education.* New York: Routledge.

Mathews, B. (2004). Promoting emotional literacy: Equity and interest in science lessons for 11–14 year olds; the 'Improving Science and Emotional Development' project. *International Journal of Science Education, 26*(3), 281–301.

CHAPTER 11

FREDERIC PERRIER

ACTIVE SCIENCE FOR CHILD REFUGEES

INTRODUCTION

Towards the end of 1798, an insurgency quelled by French troops in the Swiss village of Stans left a bloodbath with more than 300 civilians dead, including 25 children (Soëtard, 1995). This event convinced Johann Pestalozzi (1746–1827), who had come to Stans immediately, that education, and not the Revolution, was the hope for a better future. He then gathered orphans and homeless children roaming through the countryside torn by war, and he organised a home and a new educational approach based on the joy of learning. The ideas of Pestalozzi and his followers can still inspire us today. More than 200 years later, indeed, despite much technological progress and apparently sophisticated ways of life, millions of child refugees still flood all continents, victims of extreme poverty, natural disasters, war, and genocide (Yule, 2000). Sometimes children have become active participants in the conflicts and even perpetrators of atrocities (Bracken et al., 1996; Pearn, 2003; Summerfield, 1999).

Child refugees, whether met in refugee camps (e.g. Spouse, 1999) or in the streets (Williams, 1993), first of all, need emergency medical assistance (Pearn, 2003). In addition, they are at risk of developing severe psychological distress, as confirmed by detailed studies for example after the 1988 Armenian earthquake (Pynoos et al., 1993), and after massive violence in Rwanda (Dyregrov et al., 2000) or in Bosnia (Papageorgiou et al., 2000). To address this problem, remarkable psychological assistance programs have been set-up, using various therapeutic techniques such as clinical interview, story telling, or drawing (Yule, 2000). The theoretical foundations for such interventions are mostly derived from the work of Anna Freud, Melanie Klein, and Donald Winnicott, performed with child refugees in England during the Second World War, but their relevance for conflicts in developing countries in Asia or Africa is a matter of controversy (Summerfield, 1999). In particular, the model of posttraumatic stress disorder (PTSD), which encompasses clusters of recurring pathological symptoms (Yule, 2001), is questioned for non-Western cultural contexts, and especially for children.

Despite the lack of consensus on the mental health issues of child refugees, one may try to address their needs, and, in particular their education, following the footsteps of Pestalozzi. What type of education can we offer to children shorn by disaster and death? In our society overwhelmed by technology and products of scientific thoughts, what type of science education can we propose? For all children, definitely, but especially in the case of the child refugees, we expect from education more than just a school routine and an efficient programming to pass exams. We expect an education reviving

Steve Alsop (ed.) Beyond Cartesian Dualism, 137–148.

the desires of the learner and the involvement of the teacher, an education including affective values and emotions (Alsop, 2001; Rogers, 1974), in the constructivist spirit (Martin, 1999).

Active science methods, such as hands-on methods (Haury & Rillero, 1994), promote the construction of concepts from a direct acquaintance with natural phenomena, through experiments, and emphasise the joy of discovery (Lederman, 2001). Such activities therefore must be particularly valuable for child refugees. In this chapter, we discuss in a general perspective how active science can be practised with child victims of war. We propose a practical scheme for an intervention in a refugee camp and we propose guidelines using lessons learnt from pilot sessions performed in Rwanda in 2000 (Perrier & Nsengiyumva, 2003). Theoretical issues are discussed briefly in the conclusion.

GENERAL CONSIDERATIONS

The Setting

Let us consider a refugee camp set-up in the framework of an international relief campaign, where military protection guarantees some level of security. We assume that the site has been running for some weeks, and that the basic needs such as medical assistance, water, and food are more or less functioning. The considerations developed in this chapter should also apply to shelter homes or permanent refugee settlements.

Beyond the material discomfort, such refugee camps are usually affected by numerous problems (Berk, 1998), including political tensions, which tend to increase the trauma exposure of the sheltered population, in particular the children (Spouse, 1999). The prevalence of serious mental health problems, including major depression, can be of the order of 50% (De Jong et al., 2000), and often the parents, when present, are not able to take care of their children. Therefore, the children suffer from helplessness and boredom, and there is a great demand for all kinds of activities (Paardekooper et al., 1999).

The Potential Participants of the Active Science Sessions

The children found in refugee camps display vastly different behaviours (Yule, 2000). Some are affected by severe psychiatric conditions, such as severe depression or autistic states. Most children however, despite their terrible histories and resulting suffering, although eligible for diagnosis of PTSD in a large fraction of cases (Dyregrov et al., 2000), are apparently functioning, and they may benefit from educational sessions if they wish to participate. Some children could even be considered as resilient, a condition of stimulated achievement and cognitive growth (Monaghan-Blout, 1996). In some cases, such as street children or child soldiers, it is sometimes claimed that active survival constitutes a protection factor and contributes to the restoration of self-esteem. However, there is also indication to the contrary, and active survival can increase emotional problems (Qouta et al., 1995).

For our purpose, given the lack of reliable diagnosis tools (Summerfield, 1999), we avoid making any assumptions about the traumatic condition of the children. We have to be aware that they have gone through loss, probably cumulative exposure to extreme violence, which nurtured associated feeling of shame and guilt for having survived, and they are confronted with further stress in the refugee camps. However, they may or may not develop trauma, and if they do, they may exhibit psychiatric inability in coping or they may be ready for psychosocial restitution (Silove, 1999). Child refugees certainly are not regular pupils and, although they should not be *a priori* pathologized, they might display concentration, attention, and memory problems (Diehl et al., 1993), bias in attention, with selective processing of information and hypervigilance for threat (Dalgleish et al., 2001).

The encounter with such children is certainly an extremely moving experience, a situation that can lead to several pitfalls. Before the content of the active science sessions is described, it is therefore necessary to discuss the mental preparation of the team.

THE PREPARATION OF THE TEAM

Humanitarian interventions in a war-affected area involve a lot of different psychological and ethical aspects, sometimes mutually incompatible, both for the local and for the foreign team (Berk, 1998), a labyrinth of complex issues that can be referred to as a moral mine field (Kinzie & Boehnlein, 1993). In practice, the success of the program crucially depends on how the team deals with often ambiguous situations.

General Guidelines for Science Education in Refugee Camps

The general consensual guidelines that have been proposed for psychosocial interventions (Weine et al., 2002) can be followed for science education as well. The team should involve both local and foreign professionals, as well as representatives from the refugee community. The approach should be flexible and must integrate different perspectives on the subject. The foreign participation should not necessary imply foreign leadership, and in any case should be culturally considerate and sensitive. The objectives of the sessions should be debated in details before they are started, and all necessary authorisations should be secured. Also, any information disclosed by the children during the sessions should not be revealed outside the group, even to professionals, without the children's consent.

The relationship between the local and the foreign team members can be dramatically damaged by financial misunderstandings. In war-affected areas, and in former colonies, much suspicion remains deeply rooted in both sides. To avoid later difficulties, funds, when available, should be distributed in a fair manner and remuneration should be discussed openly.

A Slippery Ground for the Foreigner

Science teachers are probably less prepared than professional health workers for the situations beyond normal experience given by war and, in particular, civil war associated with ethnic cleansing and genocide, where children are not only victims but deliberate targets (Spouse, 1999). The original motivations for providing help in refugee camps may be sincere, but there is a long way from the spontaneous feeling aroused in the comfort of some developed country, behind the television screen, to the real situation in a Kigali hospital or a camp in Sudan. Burnout and collapse of volunteers are commonly observed, and provide an extra burden to the already overloaded local teams. In other cases, remaining a neutral witness may be too difficult, and expatriate personnel get involved and even indulge in deviant behaviours (Berk, 1998). Avoiding such breaches of ethics is not always easy, and a prior screening of the foreign participants in the program can be insufficient. Self-esteem and ego, if they were the only real motivation, are unlikely to survive prevailing chaos, suspicion, and, sometimes more damaging, the indifference of the setting. It will be difficult to be accepted, to win the trust of the refugees, who are no dupes and are always wary of being exploited. Videos and photos in particular should be used with utmost care, and should better be avoided in general.

In Africa, the white persons are confronted with a dual situation. On the one hand, the former colonialist keeps a somewhat special and embarrassing status, with a mixture of fascination among the smaller children. On the other hand, the white persons, for example dubbed abazungu in Rwanda, can also be identified with the cause of the suffering, and, whatever the white persons are trying to achieve, their words are not believed and their life does not weigh as much as they think.

More than the atrocities themselves, in which the foreigners are anyway rarely invited, except sometimes the press, it is the lack of clear happening, albeit with the permanent possibility of violence that will slowly challenge the mental stability of the team members. Security must be a serious matter, to be evaluated daily, but it should not become paranoia. If security of the team means sticking all the time with fellow foreigners in special four-wheel drives subject to special treatments, then it is certainly better to stay at home. Also, one should not confuse the decision of bearing some level of well-estimated risk, with anything close to courage. In the circumstances of war, courage should be left to the victims who still struggle with kindness and humanity, such as the exemplary widows of Rwanda (Baqué, 2000).

Foreigners who are involved with child refugees have also to be aware of what expects them when they are returning to their regular activities. The feelings associated with refugees, both positive and negative, leave profound scars, and "normal" life is likely to appear tasteless. The amount of subjects that can be discussed with former friends, people from "before," will tend to shrink. Loneliness can thus be a consequence that has to be accepted.

Making a Dedicated and Congruent Team

All members of the team, local and foreign, are put emotionally under test, and they all have to respond with competence, avoiding transference and countertransference vicious circles (Kinzie & Boehnlein, 1993). Nevertheless, the situations during the sessions must be dealt with a genuine feeling towards the child refugees, without avoiding empathy (Grünbaum, 1997), a situation referred to as congruence (Rogers, 1957).

Emotional challenge is permanent for the team members; intervention may often be confused with adoption, especially with the smaller children who are expressing tremendous affective demands. On the other hand, teenagers may be looking for substitutes to have conflicts with. Usually the staff of the refugee camps or shelter homes has a tough time with them, and it is comparatively easier for a punctual presence. When a trust relation is established, it can be extremely demanding, and hard to manage at times. However, the situation can change fast, the educators can be taken as scapegoats if they make a small mistake and break the trust. The safety margin between the two extreme situations is narrow.

The best predictor of success is that the team is feeling naturally at ease with the crooked situations of the refugee camps, and the participants of the team feel this adaptation rather fast, sometimes as soon as the first encounter. This situation of spontaneous comfort may be however harder to maintain after incidents have actually occurred. Basically, the team has to be genuinely happy to spend time with the refugees, just as a happy teacher is a good criterion to guarantee a successful class (Alsop, 2001). To achieve this state, and sustain it in the long run, one strategy is to prepare each session extremely carefully, with frank discussions among all members, with the feeling of the huge common responsibility shared in front of the children. Sometimes the team members may wonder whether they are not just overdoing things, but, in fact, one is never doing enough. In the hours of doubt, hard work thus can be a response.

The team has a fantastic task ahead, a path to tread together with child refugees who are in expectations, if not in demand. The investment is tremendous for the goal to be reached, and maybe the science is just the pretence of a human adventure. It should be made a great pretence anyway, and good science!

IMPLEMENTATION OF ACTIVE SCIENCE SESSIONS FOR CHILD REFUGEES

Organisation of the Sessions

After getting familiar with the setting and the community, children groups can be formed. Each group should have the supervision of two adults, one science teacher and one psychologist, preferably both speaking the language of the children, as translations lead to numerous problems (Summerfield, 1999). The number of children per group depends on the circumstances, but five is a good choice (Perrier, 2004). Larger groups can be divided in subgroups. The routine of the sessions should be at least two per week, with a duration of about 1 hour. The team should however remain available for a longer time, and be ready for more sessions if this is a request from the children. If possible, the local community and children themselves should be involved in the organisation of the sessions.

No specific room is needed to perform the activities, but some definite space should be allocated, where the equipment can be gathered and the productions can be exposed. The equipment can be kept by the participants themselves, except probably the expensive equipment such as telescope or microscope. The sessions should remain open to nonparticipants, and can even become an attraction for community members. In particular, the sessions should be open to mentally sick and mentally handicapped children, who tend to remain totally abandoned in refugee settings (Weine et al., 2002). The sessions may allow a first contact with such children, an expression of basic respect towards them, a first evaluation, and subsequent referral to a competent psychiatric team.

To make a first contact with the children and the community, it can be extremely useful to organise a special session, called the zero session (ZS) similar to a show or a game (Perrier & Nsengiyumva, 2003). This ZS can be used to introduce the topics of the scientific activities, but also to gather some basic information about the participants, such as response level or learning modalities (Martin, 1999). The main outcome of this first session should however be noise, mess, and smiles, and maybe sparkling laughs, with the aim of thawing the terrible silence and depressive apathy dominating refugee settings, and especially groups of suddenly orphaned children.

Topics of Active Science for Child Refugees

The sessions should address scientific topics that should be, first of all, "palatable" (Watts & Alsop, 1997). The main goal is to try to make sense of the meaningless, despairing past experience and present refugee environment, and convince the participants that, somehow, the world can be grasped, which makes it more beautiful and interesting to live in. Science is in essence transcultural, therefore there is no reason that the topics are matched to a particular society or conflict. The basic assumption is that all children are in essence curious and eager to learn about their world.

Examples are listed in Table 1. The name of the topics, and their attached scientific concepts are less important than the choice of a basic question, referred to hereinafter as the leitmotiv, which is a recurrent question behind the activity. The question might not be announced as such, but, by the end of the sessions, some answers and answer strategies should be the outcome. For example (Table 1), the topic "seeds and plants" may refer to a classical boring set of lessons about the anatomy of a seed, but the leitmotiv "Do plants always need light?" indicates that the main topic is an experimental comparison of germination with and without light, and of plant growth with and without light. Thus, from the leitmotiv, an inquiry-based approach (Martin, 1999) emerges immediately.

Table 1 gives only a few examples that have been tried in refugee settings. More topics can be attempted, with many different leitmotivs. The topic "insects and spiders" was found very successful (Van Cleave, 1998) and is taken as a case example in the following. The fascination with these small creatures seems to be universal, and is

Table 1. Examples of topics for active science for child refugees

Topic	*Leitmotiv*
Insects and spiders	What are differences between insects and spiders?
Dinosaurs	Where are the dinosaurs now?
Stars and planets	Where is Jupiter in the sky?
Seeds and plants	Do plants always need light?
Electricity	Can five players with each a switch control one bulb?
Electronics	Can you make an organ playing when there is light?

equally valid with all age groups, including advanced university levels, a fact that may be interesting to investigate in details (Shepardson, 2002). In the context of a refugee setting, this topic certainly guarantees that there is no shortage of the primary material, and in particular mosquitoes, flies, spiders, cockroaches, and more!

Common aspects of possible topics are that they should be feasible in the refugee setting, and that they should allow for the construction of big models. For example, the topic "insects and spiders" can lead to the fabrication of a large bamboo mosquito, or the leitmotiv "Where is Jupiter in the sky?" can lead to the construction of a 3D sky map with cardboard. Such large models decorate the setting, and play an important role in the therapeutic approach (Perrier, 2004).

Equipment

The equipment needed for experimental science is a regular source of complaint supported by the lack of funds, and the most illegitimate excuse for dismissing the activity. In fact, a minimal set of material needs to be transported to the refugee setting: cutters and extra blades, multifunction pocket knives, small screwdrivers, markers, scissors with rounded tips (many), glue (hard to find on site), strings (thick and thin), measuring tape, rulers (long), card games, blank visiting cards, LEGO™ boxes, painting brushes and lots of water colours, balloons, table tennis balls, and children science encyclopaedia. Sometimes not enough trash material is available in the refugee camps, especially in countries where recycling genius is prevalent. However, a lot of brainstorming can be generated by candles, toothpicks, matches, which are always available everywhere, and by a few old mineral water bottles and cans that can be salvaged somehow. For example, tools to catch live insects can be made by cutting a mineral polymer bottle in two halves: the upper half can be used to catch the insect against a flat surface. Then, by inserting a postcard, the prisoner can be dropped to the lower half of the bottle used as a box. Also, when the upper half is put upside down in the lower half with a banana, then the trap is full of fruitflies after a few hours. If a bulb is inserted, it is a trap for night insects. Ant labyrinths can be constructed with mineral water bottles and drinking straws.

In addition to such self-made tools, a microscope and a telescope can be acquired with a minimum financial investment, which opens tremendous possibilities for fascinating investigations. The children are certainly overjoyed by such instruments. Bringing a few such decent instruments, albeit lots of trouble, can also be considered as an expression of basic respect towards the child refugees.

Structuring Tools of the Sessions

Active science sessions require a lot of preparation, which is more than just gathering the equipment, in particular with a therapeutic perspective in difficult and often unfriendly environments. Once a topic has been selected, it is useful to rely on a solid operational framework for this preparation. One example for a set of rules and points, that can constitute the structural skeleton of the science education, is the concept of activity structural chart (ASC) detailed in Table 2.

Table 2. Activity structural chart (ASC) and application to the Topic "insects and spiders"

Activity structural chart	*Example*
Topic	Insects and spiders
Leitmotiv	What are the differences between insects and spiders?
Concept targets and concept maps	Differences between insects and arachnids (number of legs, body parts, eyes, caring of the young, moulting, and instars), food web, systematic classification of arthropods
Misconceptions hunt	Spiders are insects
Graded Questioning Grid:	
QT: Triggering (simple, obvious)	Do spiders have wings?
QKV: Knowledge/Verification (one answer)	How many eyes have spiders? How many legs have ants? How many wings have mosquitoes? How many wings have dragonflies? Do ants eat sugar? Do spiders eat mosquitoes? Do spiders fly?
QEH: formulate explanation and hypothesis (more than one answer)	What do ants eat? Can you see differences between female and male mosquitoes? Are all threads in the spider web identical?
QI Investigation: formulate and perform inquiry	Where do maggots come from? Are mites spiders or insects? How do the ants find their way back home? What happens if there are no more spiders? Why are some butterflies flying fast and some flying slowly?
Instruments	Magnifying glass, microscope
Techniques and practice	Catch live insects and spiders, build traps, ant labyrinths, feed larvae and babies
Measurements	Lifetime of mosquito larva, number of ants actually carrying goods in an ant trail, weight of insects
Open Museum Models	Models of insects and spiders, model of spider web with construction guide, masks of insect faces
Drawings	Insects and spiders body parts, insects and spiders within their ecosystem

Table 2. (continued)

Activity structural chart	*Example*
Affective dimensions	Watch birth of butterfly, mother spider carrying babies, give names to prisoner insects, set insects and spiders free again
Emotions	Do insects and spiders have blood? Afraid of spiders?
Music and oral expression	Reproduce sounds of crickets, cicadas. Make an insect noise symphony
Theatre and acting	Play insect folk tales, food web stories
Applications	Recognise malaria-carrying mosquitoes, recognise the cause of scabies and derive hygiene tips, identify dangerous species, learn what to do in case of insect bite.

The ASC includes the leitmotiv, for which several variants can be used. The underlying concepts are reviewed in the ASC and some concept targets selected for the sessions, including a concept map (Martin, 1999) and some known misconceptions to be hunted. Another important tool is a graded questioning grid (GQG). The GQG is a 4-level ladder of questions (Perrier & Nsengiyumva, 2003), ranging from obvious triggering questions (QT), to questions with an answer that can be checked (QKV), questions with several possible answers (QEH), and finally reaching questions whose answers are strategies and a complete investigation (QI). A different set of questions following the GQG can be prepared for every different session. These questions can be used during the ZS and in many different ways during the sessions. For example, the QKV examples listed in Table 2 can constitute a game during which the children, grouped in competing teams, have to find the answers themselves, using captured insects and magnifying glasses.

The ASC (Table 2) also mentions explicitly the available instruments, and techniques to be taught during the sessions, as well as 3D models and drawings. In our context, opportunities for involving emotional and affective dimensions should be listed. Finally, practical applications that may be of immediate relevance for the children and their community should emerge. It is important to give the children the feeling that they can participate in the life of the camp and provide help to their relatives (parents and siblings), a restoration of self-esteem that can contribute to resilience (Rutter, 1987).

The active science toolbox contains more tools beyond the ASC. Every educator may have his preferred messages that he may want to induce. Such messages may include: "I wonder why this and that?," "It does not work, great, let's investigate why!" or "I don't know, let's see by ourselves." Finally, during the sessions, it may be appropriate sometimes to open the "silly corner." This is a set of funny experiments, jokes, and magic tricks. During the insect sessions, the silly corner may be to paint fancy colors on the back of large cockroaches.

Management of the Sessions

The preparation and the ASC, intentionally, do not include a detailed program with a minute planning for each session. The ASC is not a schedule or a road map to be followed, rather it is to be considered, first of all, as a safety net. Beyond the preparation, the most important aspect of the session, of course, is what actually happens with the children, and whether an inquiry-based approach is actually taking place. For this purpose, many adaptations need to be introduced, likely with much improvisation. If for example a child proposes their own questions to be investigated, they become immediately the topic of the day. However, a detailed ASC contributes to a successful improvisation, because many possible questions have already been considered as options, and materials or tricks are ready.

The role of the team during the sessions remains to be assessed in the considered application by dedicated research. Watching and listening from a distance, with discretion suggested by orthodox constructivism (Martin, 1999), may be insufficient. The team has to intervene and ensure positive outcomes, sometimes by helping, sometimes by manipulating the events. Also the team has to worry about the relationship between the children and promote values of mutual respect and understanding, and to show to the children how to manipulate fragile instruments and small living creatures, thus trying to develop respectful and soft behaviours.

Beyond the professional clinical attitude, the team should first of all be sincere during the sessions and be genuinely interested by the topics and the results of the investigations performed by the children, respecting the way they are attacking the problems.

CONCLUSION AND OUTLOOK

In this chapter, I have outlined the motivations for undertaking active science with child refugees, and presented some practical guidelines. Most aspects, however, remain tentative and preliminary, and should be the subjects of dedicated future studies. An important ethical concern in psychosocial intervention is the quality of the work (Weine et al., 2002) and therefore evaluation schemes need to be implemented. For self-evaluation, tools and checklists should be designed and experimented, and prototypes developed for scientific education (i.e. Laukenmann et al., 2003) could be modified for this particular application. More problematic is the evaluation of the impact of the sessions in terms of evolution of symptoms. Indeed, the theoretical foundations for such work remain sketchy at this stage. Can experimental science indeed provide a safe scheme for psychosocial resilience? Can such sessions restore the developmental pattern of learning after trauma, leading to a successful pathway for the children of war? What areas of the unconscious are involved in active science and investigative cognition? Is it legitimate to speak of therapy?

Despite such open questions, active science can potentially play an important role for child refugees, and contribute to resilience, and the numerous possibilities offered by active science education should be explored in all settings, and certainly including refugee settings. Active science definitely offers unique roads to psychotherapy and maybe interesting alternatives, with the powerful impact of the joy of touching, joy of guessing, joy of learning, and joy of discovery. In places where suffering is often unbearable, emotions can be metamorphosed into a shared enjoyment. Scientists, science teachers, engineers, should therefore contribute at the front line, together with the doctors, the psychologists, the religious comforters, hand in hand, with their own message.

Far from being an excuse to smuggle noxious concepts of the superiority of the so-called developed Western world, active science can be a way to share with child refugees the humbling quest for an intimate relationship with natural phenomena, a vision from childhood to carry well into adult age. Education is always an adventure, and the opportunity of active science should be proposed to child refugees, an opportunity for self-reconstruction, for inner expression, an opportunity to jump back into the world and into a new future, not avoiding passion, and including, hopefully, a lot more compassion.

ACKNOWLEDGEMENTS

The author is forever grateful to the children of the Benebikira orphanage (Ruhengeri, Rwanda) and the children of the "Home of New Hopes" (Lalitpur, Nepal) for their many smiles and overwhelming enthusiasm, and he thanks Nicole and Jeeten Thakuri for their friendship and for unravelling the way into action. Catherine Crouzeix is thanked for comments on the original manuscript.

REFERENCES

Alsop, S. (2001). Seeking emotional involvement in science education: Food-chains and webs. *School Science Review, 83*, 63–68.

Baqué, S. (2000). *Dessins et destins d'enfants*. Nice : Hommes et Perspectives.

Berk, J. H. (1998). Trauma and resilience during war: A look at the children and humanitarian aid workers of Bosnia. *Psychoanalytic Review, 85*, 639–658.

Bracken, P. J., Giller, J. E., and Ssekiwanuka, J. K. (1996). The rehabilitation of child soldiers: Defining needs and appropriate responses. *Medecine Conflict Survival, 12*, 114–125.

Dalgleish, T., Moradi, A. R., Taghavi, M. R., Neshat-Doost, H. T., & Yule W. (2001). An experimental investigation of hypervigilance for threat in children and adolescents with post-traumatic stress disorder. *Psychological Medicine, 31*, 541–547.

De Jong, J. P., Scholte, W. F., Koeter, M. W. J., & Hart, A. A. M. (2000). The prevalence of mental health problems in Rwandan and Burundese refugee camps. *Acta Psychiatrica Scandinavia, 102*, 171–177.

Diehl, V. A., Zea, M. C., & Espino, C. M. (1993). Exposure to war violence, separation from parents, post-traumatic stress and cognitive functioning in Hispanic children. *Interamerican Journal of Psychology, 28*, 25–41.

Dyregrov, A., Gupta, L., Gjestad, R., & Mukanoheli, E. (2000). Trauma exposure and psychological reactions to genocide among Rwandan children. *Journal of Traumatic Stress, 13*, 3–21.

Grünbaum, L. (1997). Psychotherapy with children in refugee families who have survived torture: Containment and understanding of repetitive behaviour and play. *Journal of Child Psychotherapy, 23*, 437–452.

Haury, D. L., & Rillero, P. (1994). *Perspectives of hands-on science teaching* (ERIC ED 330 584). Columbus: The ERIC clearinghouse for science, mathematics and environmental education.

Kinzie, J. D., and Boehnlein, J. K. (1993). Psychotherapy of the victims of massive violence: Countertransference and ethical issues. *American Journal of Psychotherapy, 47*, 90–102.

Laukenmann, M., Bleicher, M., Fuss, S., Gläser-Zikuda, M., Mayring, P., & Von Rhöneck, C. (2003). An investigation of the influence of emotional factors on learning in physics instruction. *International Journal of Science Education, 25*, 489–507.

Lederman, L. (2001). Revolution in science education. *Physics Today, 54*, 11–12.
Martin, D. J. (1999). *Elementary science methods: A constructivist approach*. Wadsworth: Thomson Learning.
Monaghan-Blout, S. (1996). Re-examining assumptions about trauma and resilience: Implications for intervention. *Psychotherapy in Private Practice, 15*, 45–68.
Paardekooper, B., De Jong, J. T. V. M., & Hermanns, J. M. A. (1999). The psychological impact of war and the refugee situation on South Sudanese children in refugee camps in Northern Uganda: An exploratory study. *Journal of Child Psychology and Psychiatry, 40*, 529–536.
Papageorgiou, V., Frangou-Garunovic, A., Iordanidou, R., Yule, W., Smith, P., & Vostanis, P. (2000). War trauma and psychopathology in Bosnian refugee children. *European Child and Adolescent Psychiatry, 9*, 84–90.
Pearn, J. (2003). Children and war. *Journal of Paediatrics and Child Health, 39*, 166–172.
Perrier, F. (2004). Practising active science with child refugees: A clinical perspective. *The Science Education Review, 3*(2), 67:1–67:20.
Perrier, F., & Nsengiyumva, J.-B. (2003). Active science as a contribution to the trauma recovery process: Preliminary indications with orphans from the 1994 genocide in Rwanda. *International Journal of Science Education, 25,* 1111–1128.
Pynoos, R. S., Goenjian, A., Tashjian, M., Karakashian, M., Manjikian, R., Manoukian, G., Steinberg, A. M., & Fairbanks, L. A. (1993). Post-traumatic stress reactions in children after the 1988 Armenian earthquake. *British Journal of Psychiatry, 163*, 239–247.
Qouta, S., Punamäki, R. L., & Sarraj, E. E. (1995). The relations between traumatic experiences, activity, and cognitive and emotional responses among Palestinian children. *International Journal of Psychology, 30*, 289–304.
Rogers, C. R. (1957). The necessary and sufficient conditions of therapeutic personality change. *Journal of Consulting Psychology, 21*, 95–103.
Rogers, C. R. (1974). Can learning encompass both ideas and feelings? *Education, 95*, 103–114.
Rutter, M. (1987). Psychosocial resilience and protective mechanisms. *American Journal of Orthopsychiatry, 57*, 316–331.
Shepardson, D. P. (2002). Bugs, butterflies, and spiders: children's understanding about insects. *International Journal of Science Education, 24*, 627–643.
Silove, D. (1999). The psychosocial effects of torture, mass human rights violations, and refugee trauma: Toward an integrated conceptual framework. *Journal of Nervous and Mental Disease, 187*, 200–207.
Soëtard, M. (1995). *Pestalozzi.* Paris: Presses Universitaires de France.
Spouse, L. (1999). The trauma of being a refugee. *Medicine, Conflict and Survival, 15*, 394–403.
Summerfield, D. (1999). A critique of seven assumptions behind psychological trauma programmes in war-affected areas. *Social Science and Medicine, 48*, 1449–1462.
Van Cleave, J. (1998). *Insects and spiders.* New York: Wiley.
Watts, M., & Alsop, S. (1997). A feeling for learning: Modelling affective learning in school science. *The Curriculum Journal, 8*, 351–365.
Weine, S., Danieli, Y., Silove, D., Van Ommeren, M., Fairbank, J. A., & Saul, J. (2002). Guidelines for international training in mental health and psychosocial interventions for trauma exposed populations in clinical and community settings. *Psychiatry, 65*, 156–164.
Williams, C. (1993). Who are "street children"? A hierarchy of street use and appropriate responses. *Child Abuse & Neglect, 17*, 831–841.
Yule, W. (2000). From pogroms to "ethnic cleansing": Meeting the needs of war affected children. *Journal of Child Psychology and Psychiatry, 41*, 695–702.
Yule, W. (2001). Posttraumatic stress disorder in the general population and in children. *Journal of Clinical Psychiatry, 62*, 23–28.

CHAPTER 12

MIKE WATTS

ORCHESTRATING THE CONFLUENCE: A DISCUSSION OF SCIENCE, PASSION, AND POETRY

INTRODUCTION

Discussions of education are replete with metaphors and, as befits one on the slights of language, this discussion is no exception. A key metaphor here is that of orchestration: the teacher's role in orchestrating the teaching and learning activities taking place within a formal or a semiformal educational setting. The idea of "teacher-as-orchestral-conductor-of-learning" is not new (Watts, 2003) and, like all metaphors, has a range of convenience within which lies relevance and utility but, outside, lies a quick demise. Within the metaphor, the teacher can be seen to lead an ensemble of students, seeking to produce a composed and harmonious performance from a broad range of individual players, often working in concert. During the classroom session, the teacher looks to coordinate the contributions from various sections of the band and lead the whole to a pleasingly concordant outcome.

But here the metaphor begins to weaken: there is abundant research in science education (and education generally) to demonstrate that learners (a) do not all learn at the same rate, (b) do not all learn in the same way, and (c) do not express their learning outcomes in the same manner. That is, the teacher-conductor is not uncommonly faced with a group upwards of 20 or 30 individuals who play at different tempos—indeed, who all play variations in faintly different musical styles, on different instruments and, dispiritingly, do so with markedly varying degrees of ability, volume, confidence, and pizzazz. Not unusually, learners have quite different perceptual preferences, different motives, learning goals, and ambitions, different levels of dependence and emotional engagement with the work before them, and varying degrees of cooperation and collaboration with their peers, the learning environment, and the teacher-conductor. Seen in this light, then, the "orchestration of learning" is an idea that might have limited appeal.

But, back to the positive. Dealing with this variation in learning does require effective planning that approaches the "movements," "themes," the pace, and tempo of a lesson. An effective teacher conducting an effective lesson picks up on, and manipulates, the mood and mind-set of the learners—such a lesson almost certainly involves the emotions as well as the intellect. The orchestration of learning in this sense involves, then, bringing together the cognitive, affective, and conative (actional) dimensions of learning (Alsop & Watts, 2000). With this in mind, the teacher-conductor of science is commonly handicapped by the traditional perception of science itself as purely intellectual and quite distinctly separate from issues of emotion—a theme that forms a centrepiece to this discussion. Moreover, while learners in a school science classroom

Steve Alsop (ed.) Beyond Cartesian Dualism, 149–159.

may very well experience some interplay of thinking, feeling, and doing, it is not uncommon that this is (a) a less-than-positive experience, and (b) unlikely to have come about through design and deliberate planning. Science teachers, certainly in the United Kingdom and most probably internationally, are almost exclusively blinkered by the twin needs to "cover the content of the syllabus" and "ensure good examination grades" so that wider educational niceties are easily sidelined.

The school curriculum for science in the United Kingdom, as elsewhere, is tightly constrained—with little opportunity of portraying science as part of living culture. Millar and Osborne (1998), for example, have called for a more futuristic curriculum for the sciences in order to address the wide-felt criticism that school science overemphasises the role of a particular, and fairly narrowly defined, type of empirical practical work in communicating and understanding of the practices of science (p. 29). They suggest that, better, the science curriculum should recognise the power of "narrative stories," models, and metaphor in the teaching of science, so as to seek to foster a sense of wonder, enthusiasm, and interest in science (p. 12).

A more overt and welcome departure from the cast-iron grip of stereotypical science curricula comes from the editors of the advanced level physics (post-16) initiative for UK schools (Institute of Physics [IoP], 1999) when they say

> Scientific ideas have affected the way we all experience the world. Philosophers, poets and novelists have all been touched by them. But conventional science teaching does us all a mis-service, misrepresenting the nature of science and at the same time alienating learners. There is a great need to re-establish the human-ness of science (p. 4).

Elsewhere, the integration of intellect, emotion, and behaviour in learning has been called "confluent education." This metaphor was first introduced by Brown (1971) to signal the deliberate integration, the flowing together of the cognitive, affective, and psychomotor dimensions of learning, to be combined in a single educational experience. The intention behind confluent *science* education would be that it not only gives the learning of science greater impact and makes it more memorable but also appeals to individual learners, engages them more fully, and thereby enhances their self-esteem as learners. As Caine and Caine (1991) say

> We do not simply learn. What we learn is influenced and organised by emotions and mindsets based on expectancy, personal biases and prejudices, degrees of self-esteem, and the need for social interaction. Emotions operate on many levels, somewhat like the weather. They are ongoing, and the emotional impact of any lesson may continue to reverberate long after the specific event (p. 82).

Therefore, to mix metaphors, the role of the teacher-as-conductor of confluent science education entails orchestrating the emotional climate of science lessons in order to create a strong setting for the learning of science. As discussed here, a useful vehicle for achieving these aims is the use of poetry in science lessons.

A BRIEF HISTORY OF POETRY AND SCIENCE

Science and poetry are so often caricatured as being at opposite ends of a cultural spectrum, the one so dour, impersonal, detached, and stone-cold logic, the other emotional, irrational, imaginative, and artfully eloquent. In fact, William Wordsworth once defined poetry as the very opposite to what he called "matters of fact" (and by which he meant science). The two are, he argued, different parts of the world entirely—the merits of one are its subjective picturing and lyrical exposition of moral truths, the other its objective, precise, dispassionate cool logic and reason.

This is a debate which is part of a long and familiar quarrel pursued in Romantic literature from Wordsworth to Lawrence, a struggle between Snow's "two cultures" which is still alive today—for example, in curricula allocations, the partitioning of government funding, and the general snobberies of life. In the United Kingdom, for example, to seek a future in culture and "the Arts" has a far higher social cachet than to want to be a scientist (or, worse, an *engineer*!) with all the echoes of Samuel Taylor Coleridge's announcement that "I believe the souls of 500 Sir Isaac Newtons would go to the making of a Shakespeare or a Milton."

To poets at that time the new technical language seemed a sterile sea of jargon, in which the imagination would freeze and drown. John Donne was the first and last English poet not to feel like this about scientific language. He was fortunate, being born at just the right time—1572—after the beginning of modern science but before its specialised technical vocabularies had really taken off. So for Donne, scientific language could still be warm, mysterious, and sonorous, like poetry. He could think of love, and the scientific methods used for establishing latitude and longitude, as perfectly compatible and mutually enriching subjects. Arthur Koestler (in *The Sleepwalkers*) is adamant about the "wrong-headedness of setting up academic and social barriers" between the arts and humanities, the artificial demarcation that has since grown up and that commonly bars people from being at home in both camps. What are needed, he argues, are "creative trespassers" as a means of ending the "cold war" between the two cultures.

Recent work in this area (for example, Watts 1989, 2000; Watts and Barber 1997) has unearthed some, but not glorious amounts of scientific poetry. Considering both the enormous inventive imagination that is science and the huge human implications that it holds for people it seems strange that poets have not used it more. Carey (1995) suggests that, among the English poets even Shelley, who knew more about science than most, does not really write scientific poetry. To treat the cloud, says Carey, as a poem about meteorology—though it is that—would be to ignore most of its meaning:

> Generally speaking, science has had a bad effect on poets, inciting them to bombast (of the O thou terrestrial ball variety) or to drivelling regrets that science has banished "faery lore."

One reason for this, he suggests, is that it is commonly assumed that poetic imagination is superior to scientific imagination, so poets simply need not bother with science.

A FOCUS ON FEELING

"Coldly mechanical" remains a lingering view of science: scientists are still commonly portrayed and perceived as detached, purely intellectual, often darkly lacking in warmth or compassion. The challenge for science education, then, lies in rehumanising science, creating a better confluence of the intellectual, the emotive, and the practical.

Two routes are adopted here. The first trades on the use of image and imagination. Words and images are among the scientist's tools, they can have a "power and influence quite out of proportion to their triviality as mere marks on paper or vibrations in the air" (Sutton, 1992, p. 1). Science, says Sutton, is strewn with linguistic models, metaphors, and analogies, many in varying degrees of completeness and sophistication. As Buckley et al. (1997) points out, models and metaphors play an important role in enabling induction from experience, envisioning, understanding, describing, explaining, predicting, controlling, and prescribing all manner of phenomena. Creative, imaginative, or inventive activities, playing with words or toying with language are a full part of science. As in science, so it is in the science classroom: poetry allow learners not only to use familiar language codes and to use the registers of science but also to test their "thinking-not-yet-finished." For some teachers and learners, inventiveness and imaginative become an important factor in expressing emergent understanding and half-formed thinking (Taylor, 1997), and for expressing some of the feelings attached to that thinking.

Poetry is valuable precisely because it combines intellect and emotion, both of which have to be in balance. The poet Alan Tate (1970) has said that a good poem has "nothing to do with exalted feelings or being moved by the spirit." Rather, a poem, he says, is "simply a piece of craftsmanship, an intelligible or cognitive object" (p. 59). Matterson and Jones (2000) disagree, and seek a more rounded and inclusive form for poetry. For them, poetry is a form of communication, a way of organising conceptual and emotional reality. To write poetry is to provide a "momentary stay against confusion" (p. 165), providing some clarity and order in mood and matter for the writer.

The second route makes use of the movement between fore and ground, asking teachers and students to see particular aspects of science as coherent in themselves and yet part of a bigger, often less coherent, picture. There are times to be holist and times to be reductionist, times to be broad and general and times to be specific and particular. To observe a structure or an object holistically is to appreciate the object in its entirety and not to split off too quickly aspects of the object and reduce, dissect, measure, question, calculate, or test out theories, categorisations, or classifications. It is a well-worn cliché to say that the facts and concepts, rules and principles of school science must be seen in context but it is nevertheless a truism that bears further emphasis. There are times within science, and certainly within poetry, when we need to savour ambiguity—of doing without direct answers and working with fragmentary knowledge, bringing together the pieces of a problem-puzzle. As Bateson (1994) says "we must make do with the pleasures of recognising and playing with pattern, finding coherence in complexity, sharing within multiplicity" (p. 9). In Einstein's words

> The most beautiful experience we can have is the mysterious. It is the fundamental emotion that stands at the cradle of true art and true science. Whoever does not know it can no longer wonder, no longer marvel, is as good as dead.

The real question here is whether all of science can be conceptualised as the result of a loving and caring relationship between the known and unknown. The rehumanising of science asks for a greater transaction between personal experiential knowledge and observational spectator knowledge.

POEMS IN SCHOOL SCIENCE

Poetry in school science is not new—but it is marginal. A recent collection of poems by teachers and learners from around the world, called *Creative Trespass* (Watts, 2000), was published by the Association for Science Education. Clearly, some contributions to this anthology came from long-time closet poets, recluse in stock cupboards, waiting to "out" themselves when the climate became ripe. As one respondent wailed: "I thought no-one would ever ask!" From this collection alone it became clear that poetry is used in science classrooms as personal expression, in school corridors for display, in school competitions for awards, as mnemonic aids for physics formulae, as computer screen savers, in explorations of subtlety and holism, in theory making and laboratory processes, in metaphors and models. It is used in class and for homework, in rhymes, rhythms, raps, lines, lyrics, limericks, haikus, cinquaines, and Clerihews (to note but few) in the belief that such approaches provide pupils with a powerful means of expressing themselves and, through which, to demonstrate levels of conceptual understanding.

Route one above suggests that a prime use of poetry lies in fostering images. Route two above argues for a constant to-and-fro between the sharp analysis of science and the soft focus evaluation of poetry. Route one suggests that patterns and images can transform descriptions of the objective world. What follows is a pupil's poem:

> *It's a life-form, Jim, but not as we know it ...*
> Positively charged, waking,
> Exerting a force greater than the bed,
> 206 bones aching
> Join 200 muscles with every step, like lead.
>
> 100,000 follicles
> Persuaded, piled, parted and surrender to the
> Green polyvinyl chloride comb.
> Static . . . . crackle . . . pop!
>
> Silver tongued, silver nitrate
> Reflects and beguiles the eye,
> Pink optical apparatus scrutinise ...
> Alas! The mirror does not lie!
> (by Laurence Crummay).

This generates a mix of popular culture ("It's a life form, Jim") with science (force and follicles), with insidious questions (Are there actually 206 bones? 200 muscles?) alongside quick images (silver tongued, silver nitrate). The effect is to create an overall feeling of a "science-side" view of getting up and getting ready that is recognisable to most grumbling adolescents who "don't *do* mornings." In this sense, the main asset of poetry is that it is a distillation of experience and so shares much with science: both begin from attending to phenomena that must be accurately recorded. Both take these observations to make a synthesis that accurately represents the observations as succinctly as possible. The object of both is precise observation and interpretation. Granted, poetic observation is more than a strict research methodology and it has within it, for instance, a component of self-observation and mythical reference that adds layers of perception and perspective that are difficult to contain easily within a precise area of science. However, the argument here is that poetry and science can work together to enable learners to grow in familiarity with the concepts, facts, principles, and processes with which they are working. Such poems allow learners to use familiar language codes and to use the registers of science. Here lies the potential to make salient some of the more colourful and graphic details within science to "grab" pupils' interest and imagination. The following poem comes from Diana Syder's book *Hubble* (Syder, 1996):

Walkman.
I love
the way it tucks into my palm,
the glossy tape, the spool's perfect teeth,
the obedience of buttons.

I'm not sure what it does
but I like having
the Dolby noise reduction system, with extra bass,
the adjustments at 330 Hz, 1 and 10 kHz

and its amazing
how a twitch of my finger
makes things happen in a microchip
at a level I can't think down to.
I really enjoy being able to carry it with me,
a phase amplifier of moods on moorland strides
or plain over-the-top orchestration for a sunset
yes, I adore the music pouring straight into the middle
of my head, sluicing around meninges
and ventricles, or as if someone had stuck electrodes
right in and tickled them, causing scraps of ideas
to wriggle outwards, the excitation cascading
across synapses and the whole thing generating
a larger than life reflection that shimmers
and shifts across the sea-cave roof of my skull.
(by Diana Syder).

The evocation of feelings is clear ("I like," "I love," "I enjoy," "I adore") and these rest against the crispness of the technical terms (Dolby noise reduction, meninges, synapses). Again, the over all effect is to tangle between the awe and wonder of the science and technology with the sheer pleasure that this brings on "moorland strides" and to sunsets. The images, too, are striking: "the obedience of buttons" and the "sea-cave roof of my skull" are both worth toying with. And the next poem impersonates a well-known version with a cool scientific-eye:

If Wordsworth had been a scientist...
I wonder, only, if a cloud
Is formed on high where vapours chill,
Then all at once the droplets crowd
And, loosed from sky, as rain distil.
But lo, snowflakes conceal the trees,
Is't cold enough up there to freeze?
(by Robert Sutcliffe)

The second "poetic route" noted earlier looks to move between fore and ground—by shifting focus it is possible to view objects and ideas so that they take on a new perspective, so that learners see them in a new light:

Moondial
Sundials go out at dusk,
so I am inventing
the moondial.

This is not easy
because the moon is pale,
and much more than with the sun,
the geometry changes
as the month
the year
progresses.
But a computer calculates
the exact position for the dial,
at any given time,
is very accurate.
Like all great inventions
The moondial
Has one serious defect –
You need a torch to see it.
(by Ian Baker)

Tri bug hunting
The worm alas has none,
Man as a biped has two.
Some quadruped mammals have four,
But where is the elusive tripod –
Are they kept in cages no more?
I've seen the eight-limbed octopus,
the excessive centipede and
the over endowed millipede.
Now I've run out of places to look.
Rumour has it that there are some in the science lab,
Or do they appear in a mythical book?
(by Kate Crummay)

Both these poems, by teenagers in the middle of the usual scrimmage of school science, have the extra capacity to invert common ideas and look at them anew. Campbell (1999) codifies this as both poetry and science exploiting the unconscious mind in creating beauty, playing on the boundaries between symmetry and asymmetry, between patterns and pattern breaking. Both, he says, have their "eureka moments" when new links are made between ideas, when new metaphors are created. Both use language very carefully, though science seeks to diminish ambiguity whereas poetry uses it to enable the reader to interpret meanings. Both are quizzical of the terms and expressions we use. Below are two teachers' responses; both are their ways to encourage learners to use poetry in science lessons. In "Carbon," the verse was put on the board at the end of the topic to reinforce an idea, or perhaps as a mnemonic. The teacher presenting poetry in class, her own or favourite verses, operates as an eloquent role-model for learners and can act to restore life to science's otherwise fading image.

More metals
Brass is bold,
Steel is rolled
Gold is sold,
Bronze is tolled.

Knock your head on a girder and feel
Something is wrong if you call it *mild* steel.

Corrosion reclaims
most metals that mortals have
borrowed from nature.
(by Robert Sutcliffe).

Carbon.
From barbecues to harbour bridges,
From arcing lamps to inside fridges.
In artists' dens, in schoolboys bag,
A typist's desk, an oily rag,
A plastic sheet is on the road,
A cracker's bang is a al mode.
And a demonstration of affection,
Is nothing but a light reflection.
Above the fire and underground,
Carbon goes around and round.
(by Gill Crowther).

In this case, an attempt has been made to reach for something slightly salacious to thrill and excite and to then whet the learner's appetite for more.

A RETURN TO CONFLUENCE

The poems above have been contributed by enthusiasts and are therefore the product of a biased population: not all science teachers have a poetic bent, not all science classes are strewn with poetry, not all learners enjoy poems. Samples published in school anthologies come usually from teachers who see the virtues of using different approaches to teaching, enlivening learning, developing language skills, creating nemonics, enjoying time out, responding to different learning styles, being cross-disciplinary, being confluent.

Before waxing lyrical again about the orchestration of confluence, it is important to ask the critical questions. For example, is it possible to reconcile the scientific mindset that informs logical rationale with classroom approaches that rely on intuition and linguistic insight? Peters (1998), for example, thinks not. In teaching science, he says, it is important to impart presentational forms, methods, ways of thinking and results that are specific to the subject. Students should be introduced to scientific thinking through the way in which terms are defined, problems are illustrated and compartmentalised, hypotheses are formed, methods are tested for validity and apparent solutions are tested sceptically. Scientific thinking is targeted, logical, systematic, and conscious of its methods and cannot be brought about with the help of a "constant conversation." "How can authors and academic texts be advised to write in a personal style, not to exceed the defined density of information and to address students emotionally as well?" he asks (p. 22). Is it not the case that, with most (scientific) content, a strict objectivity of the analysis and exposition is required? Can scientists and academics really be asked to act in a user-friendly manner if their subject matter is not, in fact, user-friendly?

This is an enormously pessimistic view of scientists and severely limits the extent to which science can be made intelligible, attractive, and appealing to broad groups of nonscientists. Is it the case that, as Richard Feynman once famously said, cannot! That bringing science to the nonscientist is like explaining music to the profoundly deaf. The answer here is No. As Dawkins (1998) eloquently points out, science is far from a disinterested and dispassionate enquiry wholly lacking in any beauty or moral purpose. He argues that, in revealing the wonder of the world, both science and poetry share the ability to break through the "anaesthetic of familiarity" that surrounds us:

> Breaking through the anaesthetic of familiarity is what poets do best. It is their business. But poets, too many of them for too long, have overlooked the goldmine of inspiration offered by science (p. 15).

So, confluence implies not Peters' "didactic straightjacket" but injections of Dawkins' poetic stimulants. Shapiro (1975) describes confluent education as the "deliberate, purposive evocation of knowledge, skills, attitudes and feelings which flow together to produce wholeness in the person" (p. 119). The premise is that this then allows the learner to engage with the subjects to be learned, use imagination, take risks, focus attention, and prepare for further learning. And how does the use of poetry achieve this? Taken from the teachers' accounts in their contributions to *Creative Trespass* it is possible to divine the benefits as

Allowing learners to explore topics in science in an individualistic way
Providing positive associations with learning in a subject area where this is sometimes lacking
Using imagery as an effective means of engaging imaginative learning
Fostering and validating learners' feelings in the process of learning science
Enhancing self-awareness and personal meaning
Promoting the integration of new knowledge into the learner's world

Finally, some comments about orchestrating teaching for such "learning benefits." The film *The Dead Poets Society* portrays the actor Robin Williams who plays the role of an avant-garde teacher of poetry whose opening gambit is to instruct to his class to rip out and destroy a page of their English text. He takes fierce issue with the author of the text in suggesting that the stature and greatness of a poem can be set out scientifically as the area under a graph of importance of subject plotted against effectiveness of style. The moment in the lesson catches the imagination of the class and the mood of the pupils as they are caught between mischief, glee, and the mild misgivings about (publicly) tearing pages from their textbook. While this form of confluent orchestration fits the bill in some respects, it is unfortunate in setting scientific rationalism as the antithesis of poetic understanding.

The term "orchestration" has been used to suggest a somewhat organic approach to reaching towards curricular objectives and intended learning outcomes, relying on teachers' awareness of learners' moods and dispositions to learning, and seeking an "orchestral balance" to cater as best as possible for the greatest number in the class. As Montgomery and Groat (2002) say, "If we really want to get our message across, we need to orchestrate the "material" in a multifaceted way across the range of student learning style" (p. 2). Teachers must plan on the basis of their knowledge of students and classes, with some of their decisions being based upon technical, rational, irrational, political, and value-driven reasons. They are dealing with complex learning systems brought together in a complex teaching system. The implication of orchestration is that teachers plan the science content of their lessons and then use poems to engage their pupils through, for example

Publishing and editing a school-wide compendium ("Poems across the curriculum")
A whole-school poetry day on the theme "Time"
Songs and lyrics when learning about environmental topics or earthworms
Collaborating with the English department around particular books or poems
Writing poems about ecological principles, life cycles or organisms following field-work on a pond or a woodland
Reading poems out loud to the class when teaching about rocks and fossils

As one respondent to *Creative Trespass* commented "generally children either love or loathe them, though even the loathers are generally content to suggest a last line for a limerick". To generate confluence through poetry is to develop a well-marshalled, well–motivated, and well-orchestrated classroom—and is no easy task. It is, to adapt Cuthbert's expression (Cuthbert, 2002), to create a "mosaic on the move." But then, that is a metaphor for another time.

REFERENCES

Alsop, S., & Watts, M. (2000). Facts and feelings: Exploring the affective domain in the learning of physics. *Physics Education, 35*(2), 132–139.

Bateson, M. C. (1994). Peripheral visions. *Learning along the way.* New York: Harper Collins.

Brown, J. H. (1971). *Confluence in education: Integrating consciousness for human change.* Greenwich, CT: JAI Press.

Buckley, B. C., Boulter, C. J., & Gilbert, J. K. (1997). Towards a typology of models for science education. In *Exploring models and modelling in science and technology education from the MISTRE Group.* Reading, MA: Faculty of Education and Community Studies, University of Reading.

Caine, R. N., & Caine, G. (1991). *Making connections: Teaching and the human brain.* Alexandria, VA: Association for Supervision and Curriculum Development.

Campbell, P. (1999). *Shaping the future. Making physics connect.* London: Institute of Physics.

Carey, J. (Ed.). (1995). *The Faber Book of Science.* London: Faber and Faber.

Cuthbert, R. (2002, November). *Constructive alignment in the world of institutional management.* Paper presented to Constructive Alignment in Action, an Imaginative Curriculum Symposium by the Learning and Teaching Support Network, Centre Point Conference Centre, London.

Dawkins, R. (1998). *Unweaving the rainbow.* London: Penguin Books.

Institute of Physics. (IoP). (1999). *Post-16 initiative.* London: Institute of Physics.

Matterson, S., & Jones, D. (2000). *Studying poetry.* London: Arnold Publishers.

Millar, R., & Osborne, J. (1998). *Beyond 2000: Science education for the future.* London: King's College London.

Montgomery, S. M., & Groat, L. N. (2002). *Student learning styles and their implications for teaching.* Mimeograph, Centre for Research on Learning and Teaching, University of Michigan.

Peters, O. (1998). *Learning and teaching. Analysis and halfway from an international perspective.* London: Kogan Page.

Shapiro, S. B. (1975). Developing models of unpacking confluent education. In G. I. Brown (Ed.), *The living classroom: Innovation through confluent education and gestalt* (pp. 28–39). New York: Viking Press.

Sutton, C. (1992). *Words, science and learning.* Milton Keynes, United Kingdom: Open University Press.

Syder, D. (1996). *Hubble.* Huddersfield: Smith Doorstop Books.

Tate, A. (1970). *Essays of four decades.* London: Arnold Publishers.

Taylor, A. (1997). Learning science through creative activities. *School Science Review, 79,* 39–46.

Watts, D. M. (1989). Case Study 9: Science with rhyme and reason. In D. Bentley & M. Watts (Eds.), *Learning and teaching in school science* (pp. 102–105). Milton Keynes, United Kingdom: Open University Press.

Watts, D. M. (2000). *Creative trespass: Fusing science and poetry in the classroom.* Hatfield: Association for Science Education.

Watts, D. M. (2003). The orchestration of teaching and learning methods in science education. *Canadian Journal for Science Mathematics and Technology Education, 3*, 451–464.

Watts, D. M., & Barber, B. (1997) Poetry and science. *Primary Science Review, 50*, 7–9.

CHAPTER 13

BONNIE SHAPIRO

FROM DESPAIR TO SUCCESS: A CASE STUDY OF SUPPORT AND TRANSFORMATION IN AN ELEMENTARY SCIENCE PRACTICUM

INTRODUCTION

The student teaching experience, even under normal circumstances, is a series of problems, dilemmas, and feelings about teaching for novice teachers to work through. While the majority is successful in the teaching practicum, most encounter at least some challenges in their socialization as teacher. These challenges bring transformation in knowledge, feelings, and changing expectations of what it means to be a teacher (Aitken & Mildon, 1991; Britzman, 1991; Gunstone, et al, 1993; Johnston, 1994; Newman, 2000; Shapiro, 1991, 1996). If we are to help facilitate this transformation, studies of personal meanings that emerge for students during the experience are most worthwhile. This chapter attempts to uncover the nature of these meanings through analysis of the interplay between personal meaning and the goals of science teaching in an elementary practicum experience. The case report of Charly, a student teacher in the final teaching placement, gives details of the initial threat of failure in her final teaching practicum, then explores the resources that allowed her to achieve the dramatic transformation she accomplished to achieve success in the end.

THEORETICAL AND METHODOLOGICAL GROUNDING OF THE STUDY

The search for sources of transformation in the case study are rooted in concerns noted by Geddis and Wood (1997) who cited the ill-structured nature of the domain of teacher education practice and the incompleteness of the theory that we use to try to make sense of it (p. 624). The approach to theorizing in this report is informed by a conversational and reflective approach to the experience as the student negotiated, constructed, and reconstructed her thoughts, feelings, and teaching actions. As in similar case study research, details are presented to reveal the complexity of the study, in this case, the experience of difficulty while a student teacher. The detailing of the particular in case study work allows for insights into the general or even the universal (Simons, 1996). Details of the case contribute to understanding how we may better work with novice teachers and serve as a basis for action in the field.

An assumption about the nature of meaning central to the project is that meaning has a historical, cultural, and contextual basis. Meaning is personally and socially constructed, and because it is found through multiple perspectives, it is problematic.

Steve Alsop (ed.) Beyond Cartesian Dualism, 161–172.

Meaning is interpreted through the perspective of the one who is telling the story and the one to whom it is told (Gudmundsdottir, 1996). In this way, the research perspective emphasizes the interdependence of the individual and the learning context.

The case is concerned with issues of interpretation and meaning. It also attends to the nature of power relationships between the researcher and the researched. I established a trusting relationship with Charly as her case study tutor and field supervisor, so was not primarily in the setting as the researcher, but was the student's university field advisor. The difficulties that Charly began to encounter during the experience were unexpected and required a great deal of attention. The decision to do the research caused us to carefully consider the nature of the power relationships that existed between us. Despite her distress, Charly indicated that she felt very comfortable making her experience the focus of research and appreciated the extra time it meant that I would spend with her.

THE INTERACTION OF PERSONAL MEANING AND THE GOALS OF THE TEACHER EDUCATION PROGRAM

Very deep personal meanings emerge in the student teaching experience. A review of the literature shows very few detailed studies of personal meaning during student teaching. Outstanding exceptions are work by Britzman (1991), Johnston (1994), Geddis and Wood (1997), Darling (2000), Newman (2000), and Ayers (1993) that blend features of personal and professional meanings in learning to teach. A significant line of inquiry and analysis in the study is the literature of personal meaning in learning. Dharmadasa (1994) examined the connection between learners' metacognitive processes and the construction of personal meanings, noting "if the personal meaning of knowledge is apparent to the learner and explicitly represented to the learner's awareness during or after learning, then it influences metacognitive processes such as strategy selection, planning, organization of knowledge, progress monitoring, controlling and evaluation."

One of the most important goals of the teacher education program is the socialization of students into the teaching profession. Student teachers enter a program that has been created for them and encounter differing views about the purposes of student teaching from inhabitants of the setting: partner teachers, school administrators, faculty advisors, and the pupils they work with. If we are to provide support in these settings, we must ask questions about student teachers' *own efforts* to grow as professionals. How do students work successfully through the challenges of the practicum experience? What resources, options, and choices do they see available to them when they are having difficulty in the program? What kinds of experiences do they consider valuable in their growth? How do they describe the resources that they draw upon that make changes in thinking about teaching and learning? What is the student teacher's view of the impact of social and cultural features that mediate the experience of learning to teach? Once the practicum is complete, how do they see themselves positioned for further growth and development?

The approach to the study takes a holistic view of teaching and practical understandings surrounding the life of teaching. Such a view is situated within the developing teacher's emerging values, attitudes, and beliefs about practice as distinct from the information processing tradition of knowledge acquisition. There are two interweaving thrusts in the development of personal meaning in the practicum: (1) The development of personal meaning and identity as a teacher, and (2) The development of the practical knowledge and skills involved in learning to teach. The dynamic interplay between these areas in the student teacher's development is the focus of study because the student teacher builds knowledge of both in a holistic way that takes into account such aspects of the role of the teacher, teacher–student relationships, classroom management, and student learning. An assumption of the study is that, it is in personal experience that we ground our experience and reality (Fisher, 1991, p. 62), therefore attempts to help the student have been designed to facilitate articulation and application of her own developing ideas and beliefs about learning teaching and learning.

CONTEXT OF THE STUDY

Charly's teacher preparation experience at a mid-sized western Canadian university offers a case-based program for the development of elementary generalist teachers. Students come to the four-semester program with an undergraduate degree, or may work concurrently on a degree. The 2-year program consists of integrated coursework, tutorials, and field experiences. There are no content or subject area methods courses in the program. Rather, students work with teaching, learning, curriculum, and context issues through integrated field experiences throughout the program. They complete the final student teaching practicum experience in semester 3 of the four-semester program.

Charly was in her mid 20s and came into the teacher preparation program with a degree in General Studies. Her interest in teaching began with some very positive experiences as a coach in physical education and she also had a strong interest in art. In her final practicum experience in semester 3, Charly was placed with two partner teachers who shared the teaching of 33 grade-4 students in a high needs school. Each partner shared the responsibilities of teaching while serving the school in other capacities. Placement in this setting meant that Charly benefited from the modeling, expertise, and advice of both teachers. This also meant that Charly took direction and advice from both teachers, providing benefits, but also challenges. Charly faced difficulties in the student teaching experience that eventually reached a point of crisis. Her partner teachers were unhappy with her performance. When it became clear that Charly's difficulties were far greater than normal, I spent more time with her to provide support. Charly's partner teachers also asked for private meetings to discuss their concerns. A main source of concern was lack of classroom control. They felt she was not following their directions, and failed to complete assigned tasks or put sufficient time into planning for teaching. They complained that she was producing poor lesson plans and was not spending enough time before and after class in the school. Meetings with all parties revealed communication difficulties. While her teachers felt she was not sufficiently productive, Charly stated that she was not given enough assistance to plan lessons, to understand and administer classroom routines, or to evaluate student work. She felt often left to her own devices, without sufficient guidance. These differing expectations by both parties were further exacerbated by Charly's developing concern that the partner teachers' approach to classroom management consisted of a style of interaction that she did not believe was appropriate.

THE GREATEST CHALLENGE: TEACHING ELEMENTARY SCIENCE

The most challenging teaching area for Charly was Science. She was assigned to teach a unit titled Sound. Because of her primary interests in art and physical education, she asked if she might instead, develop unit plans in those areas. Her teachers stated that the practicum was a time to demonstrate the ability to take over a typical class, therefore required work in all subject matter areas. Charly again reported that she was not getting the assistance she needed from the partner teachers to create the unit and that she felt unprepared to teach. Her partner teachers stated she was not taking the initiative to collect the resources and materials needed to develop the unit. The partner teachers called to inform me that they wished assistance to develop a performance contract with Charly, a very serious indication that a student is in trouble. I arranged a meeting to encourage all parties to work toward better communication, asking the teachers to assist by providing guidelines and support in the development of the assigned units, and Charly to endeavor to complete the assigned tasks. Despite these meetings, the difficulties were overwhelming and Charly believed she was poised to fail the practicum. Yet, instead, by focusing on the Sound unit, she made a dramatic transformation. This story of transformation became the focus of the research project.

Charly told me that she was afraid to ask for help at times, feeling that this was an admission that she was struggling, perhaps suggesting that she was not capable of completing the task. This fear is typically felt by the student teacher as she straddles two worlds. Still a student herself, she is required to demonstrate her ability to be a teacher, fully in charge of a class. By the end of the practicum, the student is expected to demonstrate the ability to plan and implement units of study while maintaining full control of the classroom. The range of experiences and expectations for student teachers is broad. Some partner teachers will model lessons from units that they have built over many years and will provide the student teacher with extensive materials from their own resources and libraries, while others, like Charly's partner teachers expect students to develop all materials and resources completely unaided.

HOW DO I PLAN TO TEACH WHAT I DO NOT KNOW?

Charly anxiously called me to speak about the assigned Sound unit. There had been no methods courses in her teacher education program and she feared teaching science. She asked

> How do I plan to teach what I don't know myself? I'm just not scientifically oriented. I don't know enough about the topic, and like, how you break it down into pieces that the students can understand.

Neither of her partner teachers had not taught the unit before and had no resource materials to share with her. I reviewed with Charly the first steps in planning the unit. What resources and personnel were available for teaching science in the school? She found none there, so we looked at the teaching resources library at the university together. As her supervising teachers did not feel she was working sufficiently independently in planning, she felt that to ask for assistance would be seen as weakness, reflecting Britzman's discussion about the isolation of teaching (Britzman, 1991). Instead of seeing the challenge of learning to teach to be met through dialogue and support from the school community, the message of her learning environment was that the teacher must do it all herself. I met with Charly to review university resources, research studies on children's ideas about sound, sample activities, textbook materials, commercially developed units of study, and methodology text resources. We reviewed the importance of developing an overview of the subject matter taking into account the grade-level understandings of students and starting with experiences that captured the students' attention and imagination. Charly said that her greatest challenge was to organize material or "break it down into manageable units for children at this level." Several models for constructivist practice build on the premise that there are significant benefits for both children and teachers in an approach that begins with learner ideas (Driver & Oldham, 1986; Shapiro, 1994). I helped Charly with activities to find out what the children already knew about the topic, then build lessons based on their knowledge. I observed her teaching three of the science lessons. Using this approach, Charly began to have remarkable success with the subject matter presentation with students, and became deeply interested in the work. She decided to begin the unit by bringing in her violin. She gathered students onto a carpeted area of the classroom and talked about her long time interest in music. Then she played a tune for them. She asked the children to tell her about any musical instruments that they had encountered or played. As she played, she asked the children to watch the vibrations on the strings. Though usually active and distracted, the class sat entranced. In this discussion, Charly focused on the vibration of the strings stating her intention to build interest and share language and background experiences. Asoko et al. (1991) point out that ideas about sound vibration are typically limited in students' minds to objects that can be seen to vibrate. I shared the research with Charly and she extended the notion of transfer in her unit to consider structures of the ear that detect and transfer sound vibrations.

SURPRISED BY THEIR IDEAS

Charly told me that she was surprised to discover through this activity, the lack of background experience with musical instruments in the class. She shared her surprise with this, and, as the lesson progressed, began to note areas of difficulty with concepts she attempted to develop. She had the children make prediction charts both to make note of the difference between high and low sounds and also to make the distinction between frequency and pitch. When they had difficulty and confused the two ideas, she told me that this was also a confusing idea for her in elementary school and she wanted to find a way to make it clearer for students. One solution she attempted was to use the terms loud and soft versus and high and low. One of the children pointed out that he was confused about high and low because the volume control on his radio said high on one side and low on the other. Charly and I talked about the ways that everyday commonsense understanding of words used to describe scientific understandings can interfere with the grasp of concepts, as noted in research studies (Shapiro, 1994; Solomon, 1984). But what was most surprising to Charly was the difficulty many children in the class had making predictions. They did not seem to grasp what it meant to make a prediction. Some made wild guesses. We discussed the value of emphasizing that a prediction is an educated guess based on some experience we have already had with the material, or some way that we have developed to think about the outcome.

Charly shared that much of her own insight and interest in the topic came through conversations with her father, a welder. He suggested that she use bottles of water to create varying heights of columns of air to demonstrate some of the variables involved in creating differences in pitch. Her father created some materials to consider different variables. He had some extra piping at work and cut pieces with different lengths and diameters. The children were very thrilled to realize that the pipes had been cut just by him just for them. She encouraged them to predict the pitch of the pipes, but was surprised when they were reluctant to use the clues of width, length, and thickness of the pipes to make predictions. Most preferred instead, to test the pipes, by striking them. She realized, again, through working with the materials, the value of drawing their attention to variables that might affect the final tone produced.

BLENDING IDEAS ABOUT TEACHING SCIENCE CONTENT WITH APPROACHES TO CLASSROOM MANAGEMENT

Teachers' approaches to dealing with the challenge of balancing understandings of subject matter, classroom management style, and instructional practices depend on factors related to teaching contexts and the features of the learner population (Lee, 1995). The classroom management problems that Charly experienced did not disappear during the science lessons. But she built a contract with students about acceptable behavior that was very effective. At regular points during the lesson, Charly reminded the children about the contract that each had signed, agreeing to respond when given a signal. The signal, a tone Charly played on a ceramic pipe also became a focus for developing ideas in the lesson. She asked the students to comment on the pitch and loudness of the tone. Charly had originally introduced ideas about sound production in terms of strings vibrating, then in terms of water bottle and metal pipe vibration, something that students could also actually see happening. Next she began to make connections with sounds produced in another way, using materials that could not be observed to vibrate. Charly was able to build on previous activities introducing ideas about sound production, but admitted that she found the concepts difficult herself. She found that focusing on ideas that the students had difficulty understanding helped to build a deeper understanding of the subject matter. But she was also finding that an activity approach created behavioral control challenges.

NOW IS THAT WHAT A SCIENTIST WOULD DO?

Charly shared her beliefs about the importance of taking students' ideas into account, and the value of employing humor to encourage students to become aware of and comply with classroom behavior standards. In one example, students were making sounds by vibrating various lengths of string. When two students began to use their strings as slingshots, she asked them, "Now is that what a scientist would do? Do they play with their strings or work with them to learn about how they make sounds?" The children quickly changed the strings back into devices for studying sound vibrations. This successful attempt to address a classroom management issue also inserted a message about scientific attitudes and the care that scientists take in the use of materials. As Charly gained confidence she began to focus less on negative behavior and more on reminding students of the seriousness of their work as scientists.

We talked about Charly's plan to demonstrate how sounds are received through ear structures and interpreted by the brain. I shared with her the research finding that younger children associate sound reception less with the ear than older children (Watt & Russell, 1990). Children often say they are able to hear because they are listening, rather than talking about sound receiving structures. Charly, originally wanted the children to create a human representation of parts of the ear: one child representing the opening to the ear, another the eardrum, the stapes, the inner ear, etc. But because of the behavioral difficulties students were having with physical activities, she decided to develop the concept in a more formalized way, using a diagram and worksheet to focus attention. Johnston (1994) also found novice teachers moving practicum from more open-ended teaching approaches to those that more rigidly controlled student behavior. Although they then felt more successful in controlling students, they moved from their original optimistic views about how discussion and conversation should proceed in classroom interaction. Asoko (2000) reports that student teachers generally pay more attention to managing the physical rather than the mental activities of students in primary science classes, spending more time organizing activities than developing ways to help learners consider the ideas they are trying to teach. Despite the fact that student teachers described teaching in terms of subject matter topics, they spent more time on processes in teaching (pp. 88–89). The time and attention Charly spent getting to know the children proved to be very important to her development as a teacher, ultimately allowing her to work more confidently and effectively with the science content she had earlier feared. Her experience in the setting changed dramatically when she was able to discover her own knowledge about the nature of the child and build on these ideas as a foundation for classroom work. Following her initial difficulties, she took time to work with the children to clearly establish *how* they would work together, creating a much more positive environment in the classroom.

MANAGING THE FEELINGS ASSOCIATED WITH THE TRANSFORMATION

Charly's transformation involved intense cognitive and emotional demands. Charly and I found that as we spoke about her experience and about her ideas about becoming a teacher, we were talking about ideas about one's personal growth and development as a human being. It was not only the resources offered by her teacher education program that were useful in her work in the program, but also her decision to draw on the resources from her personal world that helped her successfully complete the program. During this crisis, Charly learned to clarify her beliefs about effective interaction with children, and took the risk to experiment to find approaches to meet the challenge that worked best for her and that emerged from her own closely held beliefs.

Nias (1989, 1996) writes about the importance of recognizing the emotional demands of teaching. She cites three sources of personal investment that may account for the intensity of emotions. First, the many interpersonal relationships developed in educational environments require intense interactions with students, administration, parents, and colleagues. Interaction with so many participants magnifies the emotional dimensions of the work. Another source of emotional intensity lies in the extent to which teachers invest aspects of themselves in their work. The boundaries of personal and professional identity blend in the performance of teaching acts, making the school setting an "arena for fulfillment and self-esteem." Because of this, educators are also vulnerable in these settings. They invest themselves extensively in their work and their concerns, and significant regard for their students' success and failures are intellectually, as well as emotionally intense (Nias, 1996, pp. 296–298).

Britzman (1991) writes: "learning to teach, like teaching itself, is a time when desires are rehearsed, refashioned and refused." The notion that learning to teach involves a reshaping of dreams and desires is a theme in research by Newman (2000), who studies "the dreams and goals" of preservice teachers. Newman attempted to identify, document, and follow enactments and changes in the evolving dreams and goals of six novices during their student teaching experiences. The process involves negotiating the social environments of learning. Much of the novice teacher's learning takes place within the teaching and learning culture of schools and learning about these environments includes a political dimension as well (Shapiro, et al, 1999; Shapiro, 2004).

Bell and Gilbert (1996) write that such activity requires changes in student teacher thinking and that "managing the feelings associated with the change process and viewing the change as a challenge rather than as a problem" are necessary conditions if student teachers are to be able to put new ideas into action in the classroom (p. 114). If student teachers are to make changes in their thinking, they must be offered opportunities to articulate and deal with the experiences they are having in the classroom. They note further that the change process requires "taking calculated risks, planning and knowing what else to do, having courage, confidence, control of the pace of change and ownership (119)." Not only does this require personal change, but social and political conditions that allow such change. The change process is enhanced when student teachers realize that they are not alone, that other students and practicing teachers experience many of the same kinds of thoughts and feelings.

IDENTITY, SELF-AWARENESS, AND LEARNING FROM WITHIN IN THE PRACTICUM EXPERIENCE

During the final practicum experience, student teachers seek to find their position and voice within the classroom and the community of the school, to clarify images about meaning and knowing. They achieve self-awareness (Johnston, 1994), self-identity or an emerging sense of autonomy (Kegan, 1982), or professional identity (Newman, 2000). Glazer (1999) writes that one's sense of identity can be established in two different ways: from outside in or from inside out. What comes from *outside in* is what we understand as imposition or indoctrination. What emerges from *inside out*—arises out of our experience—this we understand as expression (p. 79). Charly honored her inner wisdom and intuition and developed a creative response to the situation. Sheperd (1994) writes

> When we honor our inner wisdom, we are educated from within, the meaning of intuition. To trust our own truth is the greatest challenge in following what is most significant and meaningful to us. Our educational system and social conditioning does everything to suppress our inner knowing. When you live from your own truth, you are acting from your personal authority and you have power. The ultimate benefit of living intuitively is that you are truly yourself and all good things come from that. (p. 204)

Charly made the dramatic transformation in her teaching toward the latter part of the practicum. Her partner teachers expressed surprise and admiration for the changes and successes they observed. Charly drew on many resources to achieve this success, such as the advisement of friends and family. She also regularly sought advisement from me, her field supervisor. She struggled to establish her professional identity while retaining her personal identity. Though she was asked by her partner teachers to give them up, she remarked that she drew strength for classroom work from extracurricular pursuits, such as martial arts and running. As Britzman (1994) notes, a "repressive model of teacher identity expects teachers to shed their subjectivity to assume an objective persona. In this view, the teacher's identity and the teacher's role are synonymous (p. 92)." Charly spoke of the support, encouragement, and good humor of her family as a source of strength and creative ideas. She identified sources from within herself such as her own spiritual growth, and reiterated her ideals and values embodied in her teaching. Another powerful resource she discovered was her own intuitive sense of how to proceed when in challenging teaching events. Atkinson (2000) cites this as an overlooked and essential teaching skill.

TELLING ONE'S STORY

The research presented here gives insight into one student teacher's transformation during the teacher preparation experience, but it emphasizes the importance of seeing the teacher preparation program, not as just a place where the discussion about change takes place, but that the discussion itself is a primary vehicle for making changes in the experience. Through this conversation, both the partner teacher and the university facilitator are in key positions to help the student articulate and value changes in her own development as teacher.

Much has been written about the ways that individuals' lives develop and change through the ongoing activity of telling one's story (Bruner, 1990; Clandinin & Connolly, 1991; Connolly & Clandinin, 1986; Dixon, 1987; Fisher, 1994; Howard, 1991; Kerby, 1991; Polkinghorne, 1988, 1991; Tappan & Brown 1989). It is the telling and retelling of the stories of experience that provides continuity and coherence in one's life, but is also a vital means of reauthoring one' story and changing one's self.

In work to help students in the practicum setting, field advisors and partner teachers will benefit from creating new ways to help student teachers access their own developing worldview, images, language, and personal meaning regarding what it means to teach, and establish a clear view of the purpose and goals of the practicum. A second vital shared view is that of the roles of partner teachers and field advisors as caring guides who assist the student to learn and grow. This work suggests the importance of conversation as the foundational organizing metaphor of the experience. Case reports such as this can serve as a useful catalyst for discussion. This research is an effort to make available features of Charly's transformation and expressed hope that others will learn through her story how to help students grow as teachers. Westerhoff (1987) writes: "We are at our best when we make our lives and our search for meaning available as a resource for another's learning. To be a teacher means more than to be a professional who possesses knowledge and skills. It is to have the courage to enter into a common search with others" (p. 191).

FINAL REMARKS

Typically the university field advisor's time and involvement is more intensive with students whose encounters in the experience are troubled and this was true in the work with Charly. Because of the significant investment of time by all parties, Charly's crisis became increasingly more transparent, as more information, details, and solutions were sought throughout the practicum. Yet, the issues and perplexities encountered in this case are very similar to those faced by many novice teachers working with elementary science materials. Because of the difficulties Charly experienced, helping her involved the articulation of some of her most deeply held beliefs and conceptions about the nature of children, learning, and teaching. The case also documents sources of support and efforts Charly made to think differently about what she was doing as she moved from despair to success in the elementary science practicum experience. In summary, conversations and telling one's story are foundational metaphors for growth in the student teaching experience, not only for the student teacher's personal and social construction of the role of science teacher, but for those who facilitate professional development and growth. This work suggests the importance of being aware that novice teachers take actions of their own to bring about their own transformations. The research attempts to help develop a deeper knowledge of personal sources of insight and transformation in teacher growth so that we might help partner teachers and university supervisors become more aware of the sources that inspire and enhance their growth and development.

REFERENCES

Aitken, J. L., & Mildon, D. (1991). The dynamics of personal knowledge and teacher education. *Curriculum Inquiry, 21*(2), 141–162.

Asoko, H. (2000). Learning to teach science in the primary school. In R. Millar, J. Leach, & J. Osborne (Eds.), *Improving science education: The contribution of research* (pp. 79–93). Philadelphia: Open University Press.

Asoko, H., Leach, J., & Scott, P. (1991) *A study of students' understanding of sound 5–16 as an example of action research*. Paper prepared for the Symposium, 'Developing Students' Understanding in Science,' at the Annual Conference of the British Educational Research Association at Roehampton Institute, London.

Atkinson, T. (2000). Learning to teach: Intuitive skills and reasoned objectivity. In T. Atkinson & G. Claxton (Eds.), *The intuitive practitioner* (pp. 69–83). Philadelphia: Open University Press.

Ayers, W. (1993). *To teach: The journey of a teacher*. New York: Teachers College Press.

Bell, B., & Gilbert, J. (1996). *Teacher development: A model from science education.* Washington, DC: Falmer Press.

Britzman, D. (1991). *Practice makes practice.* Albany: State University of New York Press.

Bruner (1990). *Acts of meaning*. Cambridge, MA: Harvard University Press.

Clandinin, J., & Connolly, M. (1991). *Teacher education as narrative inquiry: Narrative inquiry as teacher education*. Paper presented at the Annual Meeting of the Australian Teachers Education Association, Adelaide, Australia.

Connolly, M., & Clandinin, J. (1986). On narrative method, personal philosophy and narrative Unities in the story of teaching. *Journal of Research in Science Teaching, 23*, 293–310.

Darling, L. F. (2000). Portfolio as practice: The narratives of emerging teachers. *Teaching and Teacher Education, 17*, 107–121.

Dharmadasa, K. (1994). *Personal meaning as a metacognitive process.* Paper presented at the Mid- South Educational Research Association, November, Nashville, Tennessee.

Dixon, A. (1987). Storyboxes—supporting the case for narrative in the primary school. *Cambridge Journal of Education, 17*, 151–156.

Driver, R., & Oldham, V. (1986). A constructivist approach to curriculum development in science. *Studies in Science Education, 13*, 105–122.

Fisher, W. (1984). The narrative paradigm: An elaboration. *Communication Monographs, 52*, 347–367.

Fisher, D. V. (1991). *An introduction to constructivism for social workers.* New York: Praeger.

Geddis, A. N., & Wood, E. (1997). Transforming subject matter and managing dilemmas: A case study in teacher education. *Teaching and Teacher Education, 13*(6), 611–626.

Glazer, S. (Ed.). (1999). *The heart of learning: Spirituality in education.* New York: Penguin Putnam.

Gudmundsdottir, S. (1996). The teller, the tale, and the one being told: The narrative nature of the research interview. *Curriculum Inquiry, 26*(3), 293–306.

Gunstone, R., Slattery, M., Baird, J., & Northfield, J. (1993). A case study exploration of development in preservice science teachers. *Science Education, 77*(1), 47–73.

Howard, G. (1991). Culture tales: A narrative approach to thinking, cross-cultural psychology and psychotherapy. *American Psychologist, 46*, 187–197.

Johnston, S. (1994). Conversations with student teachers—enhancing the dialogue of learning to teach. *Teaching and Teacher Education, 10*(1), 71–82.

Kegan, R. (1982). *The evolving self: Problems and process in human development.* Cambridge, MA: Harvard University Press.

Kerby, A. (1991). *Narrative and the self.* Bloomington: Indiana University Press.

Lee, O. (1995). Subject matter knowledge, classroom management, and instructional practices in middle school science classrooms. *Journal of Research in Science Teaching, 32*(4), 423–440.

Newman, C. S. (2000). Seeds of professional development in pre-service teachers: A study of their dreams and goals. *International Journal of Educational Research, 33*, 123–217.

Nias, J. (1989). Teaching and the self. In M. L. Holly & S. Mcloughlin (Eds.), *Perspectives on teacher professional development* (pp. 155–171). London: Falmer Press.

Nias, J. (1996). Thinking about feeling: The emotions in teaching. *Cambridge Journal of Education, 26*(3), 293–306.

Polkinghorne, D. (1988). *Narrative knowing and the human sciences*. New York: State University of New York Press.

Shapiro, B. L. (1991). A collaborative approach to help novice science teachers reflect on changes in their construction of the role of science teacher. *The Alberta Journal of Educational Research, XXXVII*(2), 119–132.

Shapiro, B. (1994). *What children bring to light: A constructivist perspective on children's learning in science.* New York: Teachers College Press.

Shapiro, B. L. (1996). A case study of change in elementary student teacher thinking during an independent investigation in science: Learning about the face of science that does not yet know. *Science Education, 80*(5), 535–560.

Shapiro, B. L. (2004). Awakening to objects of meaning in science classrooms. In T. Koballa & D. Tippins (Eds.), *Cases in middle and secondary science education: The promise and dilemmas* (pp. 119–126). Upper Saddle River, NJ: Pearson Education.

Shapiro, B., Richards, L., Ross, N., and Kendal-Knitter, K. (1999). Time and the environments of schooling, in Learning Environments Research—An International Journal, *2*, 1–19.
Sheperd, L. (1993). *Lifting the veil: The feminine face of science.* Boston: Shambala.
Simons, H. (1996). The paradox of case study. *Cambridge Journal of Education, 26*(2), 225–240.
Solomon, J. (1984). Prompts, cues and discrimination: The utilization of two separate knowledge systems. *European Journal of Science Education, 6*(3), 277–284.
Tappan, M. B., & Brown, L. M. (1989). Stories told and lessons learned: Toward a narrative approach to moral development and moral education. *Harvard Educational Review, 59*, 182–205.
Westerhoff, J. H. (1987). The teacher as pilgrim. In F. S. Bolin, & J. F. McConnell (Eds.), *Teacher renewal: Professional issues, personal choices* (pp. 190–201). New York: Teachers College Press.
Watt, D., & Russell, T. (1990). *Sound. Primary science process and concept exploration.* Project Research Report. Liverpool: Liverpool University Press.

CHAPTER 14

BRIAN MATTHEWS

EMOTIONAL DEVELOPMENT, SCIENCE AND CO-EDUCATION

INTRODUCTION

Science lessons can provide an excellent opportunity to challenge traditional stereotypes and raise pupils' emotional awareness. If one can engage boys' and girls' emotions in genuine discussions about science it can help challenge stereotypes, and so co-educational schools are an ideal place for such work. This chapter will show, using some original research, how this can be achieved.

There has always been an emphasis on cognitive outcomes in education, but Bloom et al. (1964) suggested that educational objectives could be grouped into three domains, one of which was the affective. Greenhalgh (1994) studied the connection between the emotions and learning. He argued that

> Effective learning is dependent upon emotional growth. If we are to facilitate better learning then we need to understand better the relationship between affect and learning. (p. 21)

Gardner's influential work on multiple intelligences (Gardner, 1983) continued the focus on the individual while extending the affective to include more social aspects. Gardner differentiated between two types of social intelligence, the intrapersonal, which is about access to one's own feelings, and the interpersonal. If pupils can develop emotionally it will involve being able to empathise with others. This, in turn, would enable pupils to listen and to learn from their classmates. A realisation that the development of the emotions is important has led some educators to champion the introduction of emotional literacy into schools (Sharp, 2001).

WHAT IS EMOTIONAL LITERACY?

There are many views on what constitutes emotional literacy. A central point is that while emotions may be experienced individually, they arise out of social situations and interaction with others. Hence, emotions should be seen as a response to a communal situation. Claude Steiner (1997) says: "Emotional Literacy is made up of...: the ability to understand your emotions, the ability to listen to others and empathise with their emotions, and the ability to express emotions productively. To be emotionally literate is to be able to handle emotions in a way that improves your personal power and improves the quality of life around you. Emotional literacy improves relationships, creates loving possibilities between people, makes co-operative work possible, and facilitates the feeling of community" (1997, p. 11). Steiner's emphasis uses the

Steve Alsop (ed.) Beyond Cartesian Dualism, 173–186.

emotions in relationships. In order to do this one has to focus on one's self and work to change psychologically, develop self-confidence, and become more able to manage emotions. This, in turn, can make one a better learner.

The term "emotional intelligence" is also widely used. This can have an emphasis different from "emotional literacy." For example, Salovey and Sluyter (1997) define it as "the ability to perceive emotions, to access and generate emotions so as to assist thought, to understand emotions and emotional knowledge, and to reflectively regulate emotions so as to promote emotional and intellectual growth" (p. 5). They expand on this definition, but one can see that it is basically individualistic in that the words used can be read as if they apply just to the individual. There are no direct references to social interactions or empathy with others—this is only implied. This can be contrasted with Steiner's definition, which centrally involves interactions with others.

The difference in emphasis between the terms "literacy" and "intelligence" has important implications. Using the term "emotional intelligence" has strong parallels with the way "cognitive intelligence" is used. Tests for intelligence (IQ tests) have been strongly criticised (Block & Dworkin, 1977). These tests measure a small range of attributes and infer that intelligence is property of the individual alone. Similarly, some people believe that the emotions can be quantified and measured through written tests to measure a person's emotional quotient (EQ) (Cooper & Sawaf, 1997). Written "tests" can have some meaning but the idea of EQ is a psychological approach that has much in common with IQ and rests on similar assumptions. This is an approach that I believe should be avoided, especially if used as a predictor. However, the terms "emotional literacy" and "emotional intelligence" are not used consistently. In this chapter the term "emotional literacy" will be used to try to draw attention to the social context in which the individual feels emotions.

IMPLICATIONS FOR THE CLASSROOM

The implication of the above social dimensions for this research is that teachers would benefit by being aware of, and engaging, pupils' emotions in the classroom. However, there are different degrees to which this can occur. It is possible to find out how pupils feel *about* science or science lessons, but this does not engage pupils' emotions during science lessons. This can be done through the content of science lessons (Head, 1985). For example, Alsop and Watts (2000) developed a section of work around radioactivity; its uses and dangers, so that pupils would engage their opinions and emotions, and argued that

> What we do not need is sanitised, antiseptic science but an appropriate balance of informed excitement and animated understanding. Our work continues to explore the relationship between cognition and emotion. (p. 138)

As many people have pointed out, through discussing ethical and moral issues in science lessons the emotions can be brought into play (Association for Science Education, 2002; Levinson & Reiss, 2003). The nature of discussion in the classroom about issues on which there is no agreement holds many opportunities for pupils to combine thinking with feeling. A sense of personal involvement is central to emotional engagement (Head, 1985).

In the discussion so far the pupils have only been engaged with their emotions about things. However, the work of Steiner (1997) and Greenhalgh (1994) indicates that it would be very productive to go further. Rather than just engaging pupils' emotions in order to promote learning and interest in science, we could also develop the pupils socially and emotionally, especially in co-educational settings.

This point is worth emphasising: we should explore ways of encouraging pupils to engage their emotions about each other and science to improve their social and emotional skills in general and to explore how this affects their attitudes to science.

ENGAGING PUPILS' EMOTIONS TO HELP THEM MATURE

In order to help secondary pupils to develop their social and emotional skills it is important to situate them within the context of normal subject schooling, as this is where pupils spend most of their time. One aspect of promoting the development of cognitive, social, and emotional areas is to acknowledge that they are intertwined, and to find ways of enhancing progress. Science is a natural area for such work as it is an inherently social activity and incorporates imagination, creativity, and social and political values. Developing social and emotional understanding in science lessons could give distinct advantages: (i) greater interest in, and understanding of, the nature of science, (ii) developing positive methods of communication and so getting along with people, and feeling good, as well as (iii) maintaining academic success.

For pupils to progress emotionally they need to gain an understanding of each other, and in particular, to do so across the gender divide. On this basis co-educational schools, where the other sex is present to talk to, provide the greater chance to enhance social and emotional development. This is because it is possible to engage pupils in their emotions, rather than them just being told about them and how they should change. Hence dialogue and the ensuing interplay is seen as central to helping pupils develop their sense of "self" and "other."

In an attempt to put these ideas into practice the *Improving Science and Emotional Development* (ISED) project was developed. This involved three teachers in two London comprehensive schools using three research classes and three control classes. In the research intervention group there were 82 pupils in three year 7 (ages 11–12) classes (45 boys and 37 girls). In the control group there were three parallel year 7 classes, with 46 boys and 37 girls.

In this chapter space permits only outline details of the procedures and focus on the results. For full details see Matthews (2003).

PRINCIPLES OF THE RESEARCH

For pupils to develop they need to be able to;

1. communicate with each other in a safe environment;
2. talk and listen to each other in a group situation;
3. verbalise (writing and talking) what they think went on in the interactions;
4. compare this with what other people thought had gone on (understand that there are different perceptions of the same discourse);

5. think about the social processes that went on in interactions;
6. discuss these so that they come to understand their own and each other's viewpoints;
7. empathise with each other.

Hence they will be encouraged to engage their emotions in a social setting, and this can contribute to their empathising and recognising that boys and girls bring a full range of ideas and traits to situations and can change with time.

CLASSROOM IMPLEMENTATION

In the research two boys and two girls, aged about 12 years, were put into a group of four, with another pupil acting as an observer. The group was then asked to collaboratively complete a science task (1 and 2 above). While they were doing the task the pupil-observer filled in the Discussion Assessment sheet (Appendix), which recorded how much each pupil talked, listened, and was supportive. Space was also provided for comments. When the task was finished each pupil filled in a Guesses sheet, where they had to fill in their estimates of what had happened in the group (3) Finally, the pupil-observer collected in all the sheets and ran a discussion (4 and 5), focusing on how well they had done, and any discrepancies in the estimates. Hence each pupil was, to a degree, confronted with other's perceptions of the group interactions (6 and hopefully 7).

When the pupils were used to these strategies the observer was withdrawn and about every 3 weeks they were given an Opinion sheet to fill in. These sheets required the pupils to fill in similar information to that above, but additionally asked questions like how they were finding working with the other sex? What they thought of science lessons? How much they were able to speak?

Additionally, two questionnaires with Likert-type questions were given to the research and control groups at the beginning and end of the research period to determine any changes that may have occurred in the pupil responses. One was called the "survey," which was designed to elicit their feelings about each other, science lessons, and science. The other was called "How are you feeling?" and asked questions to find out more on the groups' self-esteem and "emotional literacy" to see if the results would support the findings from the classroom work. It made no attempt to study if any individual had changed their responses over the year as written questionnaires alone cannot indicate a person's emotional literacy. Full details of the procedures and statistical analysis are given in Matthews (2003).

RESULTS

Attitudes to Science Lessons

I will start by looking briefly at the way pupils reported that their attitudes to science lessons had changed over the year. The pupils were asked three questions on how much they liked science lessons in the survey. The questionnaire used an 8-point Likert scale and the pupils' responses were entered into SPSS to establish the statistical significance (Table 1).

Table 1. *Change in pupil's attitude to science lessons*

Questions from the survey	*Research groups*			*Control groups*		
	Mean score out of 8 at start of year	*Mean score out of 8 at end of year*	*Change over the year*	*Mean score out of 8 at start of year*	*Mean score out of 8 at end of year*	*Change over the year*
How happy are you when you see science on the time table?	5.7	6.1	0.4* Boys: 0.2 Girls: 0.6*	5.9	5.4	−0.5* Boys: −0.6* Girls: −0.4#
Are science lessons interesting or boring?	5.9	6.3	0.4# Boys: 0.1 Girls: 0.7*	6.3	5.8	−0.5* Boys: −0.6* Girls: −0.3

*A matched t-test was applied: $p < 0.05$ is indicated by *; $0.1 > p > 0.05$ by #; and $p > 0.1$ is left blank.*

In Britain it is common for the pupils enjoyment of science lessons to decrease as they start secondary schools (age 12 years). The responses of the control group reflect this general trend. Hence it is satisfying that the research group showed an increase in liking for science lessons over the year and it is significant that the girls changed more than the boys.

At the end of the year the pupils were asked if they would like to continue with science if they had a choice. The results are shown in Table 2.

Table 2. Continuing with science

	Percentage that indicate that they are likely to continue with science	
	Boys	*Girls*
Research	85%	85%
Control	71%	76%

These results indicate that the research interventions were successful to a degree. To understand why this might have been we need first to look for changes in their affective abilities to see if they might be related to these changes.

Learning About Each Other

The research provided evidence on how the pupils had changed in their affective attitudes. A range of indicators can be used to gauge emotional development. In this research we focused on their ability to understand and get on with each other, and support each other in learning.

Some indication of how the majority of pupils felt doing the monitored group work is given by the quotations below:

A boy wrote in an Opinion sheet: "[Group work] makes me realise that working with the group is fun. [It makes science more interesting] because everyone helps each other." Other comments from written sheets include

> When I work with boys I begin to understand how they feel. (girl)
> I think people don't like other people because of what they see. Doing group work can make them see what the other person can do. (boy)
> They help me when I'm scared to ask [the teacher]. (girl)
> I don't look forward to working with boys, but enjoy it when I do, its better in mixed-sex groups. (girl)
> I thought that Neil was horrible, but after working with him I like him a bit more. (girl)
> I like working in a group because you learn different things from different people. (boy)
> It helps us get to know people and learn how to get along with other people. (girl)
> When you're their partner then you get to know them more and so you become better friends. (boy)

These quotations illustrate how pupils learnt about each other and began to be more relaxed with perceived differences. Evidence from the survey supports the contention that the pupils gained a better understanding of each other and gained in confidence (Table 3).

In the first three questions the research groups have shown a greater positive change than the control groups. The reason for the control group having a greater change than the research in the last question could possibly be due to the fact that the control group usually chose who they worked with, and these were usually of the same sex. Hence they had a greater opportunity to gain confidence in the same-sex situation, while the research group were put in non-friendship groups with classmates of both sexes and had to contend with pupils they did not initially get on with.

The pupils of both sexes indicated overwhelmingly that they thought that it was important that they learnt to get on with each other, and that as they did this sort of group work their understanding increased with time (Matthews, 2003).

Evidence that the research groups were at least beginning to develop realistic and grounded views of each other is supported by their responses to questions about how much the other sex would like them (Table 4).

It is possible that the last result is an indication that the pupils became more skilled at supporting and helping each other learn. Also, if the pupils did not feel that they were becoming more emotionally comfortable with the other sex they would have been more likely to respond negatively to both questions. Here are some quotations from pupils about them helping each other:

> Well, sometimes we explain to each other things, but we can understand them better than if the teacher did. (boy)
> ... you get to hear what other people wanted to say. And you learn more. (boy)
> ... ideally people will discuss it and the ones who understand one thing will explain it to the others and visa versa. (girl)
> ... you expand your memory and learn more and find out stuff what you did not know. (girl)
> Because you work with different people you can learn more, also important because you can get a better job. (boy)

We can still see that these conversations, referring to specific learning situations, are not often infused with simple statements on gender, like “girls help, but boys don’t.” This implies that the boys and girls are involved in discourses that enable them to forge gender identities that cross-traditional stereotypes and could help generate a sense of community through sharing, valuing and respecting. Of course, some pupils did not engage in such discourses, but they were in the minority.

Other confirming evidence that pupils support each other came from “How are you feeling?” which was inspected to see if evidence from a questionnaire would conflict with or support that obtained from more subjective short answers, interviews and teacher impressions (Table 5).

In each case the research groups indicated that they felt they could get support from their classmates to a greater degree than the control groups.

People who are secure in themselves can work well with a range of people. Hence, an increased preference for working in groups and with the other sex can also be taken as an indication of a “more secure” sense of self (Table 6).

Table 3.

Questions from the survey	*Research groups*			*Control groups*		
	Mean score out of 8 at start of year	*Mean score out of 8 at end of year*	*Change over the year*	*Mean score out of 8 at start of year*	*Mean score out of 8 at end of year*	*Change over the year*
I understand the *other* sex very well	5.2	5.9	0.7* Boys: 0.7 Girls: 0.7	5.1	5.4	0.4# Boys: 0.3 Girls: 0.4
I understand the *same* sex very well	7.2	7.4	0.2 Boys: 0.3 Girls: –0.1	7.6	7.67	0.0 Boys: 0.0 Girls: 0.0
I am confident of saying what I want when working with the *other* sex	5.2	6.2	1.0* Boys: 0.8 Girls: 1.3	5.3	6.27	0.9* Boys: 0.7 Girls: 1.1
I am confident of saying what I want when working with the *same* sex	6.6	7.0	0.4# Boys: 0.3 Girls: 0.4	6.4	7.0	0.6* Boys: 0.5 Girls: 0.8

*$p < 0.05$ is indicated by *; $0.1 > p > 0.05$ by #; and $p > 0.1$ is left blank.*

Table 4. Understanding and confidence with the other and same sex

	Research groups			*Control groups*		
Questions from the survey	*Mean score out of 8 at start of year*	*Mean score out of 8 at end of year*	*Change over the year*	*Mean score out of 8 at start of year*	*Mean score out of 8 at end of year*	*Change over the year*
The other sex will like you	4.7	5.0	$0.3^{\#}$ Boys: 0.2 Girls: $0.5^{\#}$	4.9	5.1	0.2 Boys: 0.0 Girls: 0.4
The other sex will help you learn	4.3	4.9	0.6* Boys: $0.6^{\#}$ Girls: 0.4	4.4	4.4	0.0 Boys: −0.2 Girls: 0.2

*$p < 0.05$ is indicated by *; $0.1 > p > 0.05$ by $^{\#}$; and $p > 0.1$ is left blank.*

Table 5. Support from friends

Questions to research and control groups	*Score at the beginning of the year, mean out of a score of 6*	*Score at the end of the year, mean out of a score of 6*	*Change over the year*	*Overall % change*
My friends in the class help me when I am upset	3.9	4.2	0.3# Boys: 0.3 Girls: 0.2	5%
Control group result	4.3	4.3	0	0%
I am able to ask for help	3.4	3.8	0.4* Boys: 0.6 Girls: 0.2	7%
Control group result	3.1	3.0	−0.1	−2%
I can ask for help and the pupils give it to me	4.0	4.2	0.2 Boys: 0.2 Girls: 0.1	3%
Control group result	4.4	4.3	−0.1	−2%

*$p < 0.05$ is indicated by *; $0.1 > p > 0.05$ by #; and $p > 0.1$ is left blank.*

The above changes are small, but indicate an increase in emotional development. However, there were also pupils who did not change, and some that indicated a strong preference for working by themselves:

> I did not like working in groups, any of it. (boy)
> I work much better by myself, the others bother me. (girl)

The Link Between Affective Development and Science

Pupil discussions enabled them to voice their feelings on (a) their learning, and (b) each other. This contributed to the affective development described above. The increase in positive attitudes to science, especially for the girls, can be linked through the increased social nature of the science classroom. The mutual support gained from the groups enabled some pupils to learn more, as other research indicates (Kutnick & Rogers, 1994).

The pupils themselves indicated these links, as indicated by the following quotations:

> It [group work] makes science more interesting and I started to look forward to coming to science. (girl)
> Group work encourages more pupils to take up science because other people can help you learn and a lot more as you get to know them. (boy)
> Science is enjoyable and it's fun when you work with people in groups, you socialise a lot. (boy)
> I would say that science is a really good subject to do, especially when you're working in groups with other people because you get to know other people and do your practicals. (girl)

Table 6. *Preferences about working in groups*

Questions from the survey	*Research groups*			*Control groups*		
	Mean score out of 8 at start of year	*Mean score out of 8 at end of year*	*Change over the year*	*Mean score out of 8 at start of year*	*Mean score out of 8 at end of year*	*Change over the year*
I prefer to work in a group rather than on my own	6.0	6.1	0.1 Boys: 0.1 Girls: −0.2	6.4	6.4	0.0 Boys: 0.3 Girls: −0.4
I prefer to work in a mixed-sex group rather than single-sex	5.8	6.0	0.2 Boys: 0.2 Girls: 0.2	5.8	5.6	−0.2 Boys: −0.1 Girls: −0.3

Table 7. Group work and science

	Negative replies (%)	*Neutral replies (%)*	*Positive replies (%)*
Does group work affect how you feel about science?	18% b = 13%; g = 23%	14% b = 5%; g = 23%	68% b = 83%; g = 54%
Would doing regular group work make pupils more interested in science?	11%	20%	69% b = 71%, g = 67%
Is group work a good way to learn science?	0%	11%	89% b = 89%; g = 89%

Note: b stands for percentage of boys; g stands for percentage of girls.

Further evidence on why the research groups may be more interested in science lessons comes from the classroom-based Opinion sheets. Three questions were directly relevant. These were (Table 7)

With the exception of the first question the boys' and girls' replies were similar in percentage terms. Here are some of their answers from the first two questions:

> Because you can compare answers and if you are right it will make you feel good. (girl)
> Because you can work with feelings. (boy)
> You get ideas more easily, so yes it can encourage pupils to take up the sciences. (girl)
> Yes, you learn from each other and it makes science interesting. (boy)
> No, 'cos we argue too much. (boy)

The pupils made the point that a major factor in how much they enjoy group work was the group they were in and how well they got on with those particular pupils.

CONCLUSION

There is evidence that emotional literacy is needed so that the intellect can be fully utilised (Goleman, 1996; Sharp, 2001). Additionally, the incorporation of emotional literacy into the school science curriculum holds out the potential for pupils to relate to each other in a more co-operative fashion, including across the gender divide, and to see science as a more social and interesting subject. This small-scale research provides evidence to support the thesis that pupils can be helped to develop emotionally. They can gain a greater understanding of each other and learn to help each other in their studies. This, when combined with the social aspects of learning in groups, can encourage pupils to take up science as a subject. In this study the amount of affective input into teaching and learning was fairly small as it was overlaid onto a National Curriculum that is heavily laden with content knowledge. If more space were to be made for the inclusion of social and ethical issues there is every reason to believe that greater affective changes could occur. As a result, it is possible that boys and girls would get on with each other better, and grow up to become men and women who would have improved relationships. Hence the results of this research provide some support for co-educational schools. Also, since pupils have to be involved in dialogue, this can also be used to improve social relationships across pupils from different cultures (Singh, 2001) as the same principles are used in the classroom as those used for boys and girls (Matthews, 2001; Matthews & Sweeney, 1997). However, for this to occur our education system needs to change (Matthews, 2005).

Even though the techniques described can contribute to science being more interesting, it is vital to remember that emotional literacy is important in its own right. Developing the affective domain could in turn help produce more scientists of both sexes who are more emotionally and socially developed. One never knows, but they may become scientists who are more socially aware and ethically responsible. And of course, an essential part of citizenship is to have the empathy to understand each other and so contribute to democracy.

For a free copy of the full report, *Improving Science and Emotional Development*, please E-mail: b.matthews@gold.ac.uk giving your full postal address.

REFERENCES

Alsop, S., & Watts, M. (2000). Facts and feelings: Exploring the affective domain in the learning of physics. *Physics Education, 35*(2), 132–138.

Association for Science Education. (2002). Can we; should we? On *ASE Science Year Resources. Bringing Science Year into the Classroom. CD-Rom.* Hatfield: Association for Science Education.

Block, N. J., & Dworkin, G. (1977). *The I.Q. controversy: Critical readings.* London: Quartet Books.

Bloom, B. S., Krathwohl, D. R., & Masia, B. B. (1964). *Taxonomy of educational objecitves. The classification of educational goals. Handbook II: Affective domain.* London: Longman.

Cooper, R., & Sawaf, A. (1997). *Executive EQ: Emotional intelligence in business.* London: Orion Business Books.

Gardner, H. (1983). *Frames of mind: The theory of multiple intelligences.* New York: Basic Books.

Goleman, D. (1996). *Emotional intelligence. Why it can matter more than IQ.* London: Bloomsbury.

Greenhalgh, P. (1994). *Emotional growth and learning.* London: Routledge.

Head, J. (1985). *The personal response to science.* Cambridge: Cambridge University Press.

Kutnick, P., & Rogers, C. (Eds.). (1994). *Groups in schools.* London: Cassell.

Levinson, R., & Reiss, M. (Eds.). (2003). *Key issues in bioethics: A guide for teachers.* London: Taylor & Francis.

Matthews, B. (2001). *Emotional literacy in secondary schools: Enabling pupils to develop in subject lessons.* Retrieved 2002, (accessed April 2002), from http://www.nelig.com/articles/ised.htm

Matthews, B. (2003). *Improving science and emotional development* (*The ISED project*). *Emotional Literacy, citizenship, science and equity* (2nd ed.). London: Goldsmiths.

Matthews, B. (2005). Engaging Education. Developing Emotional Literacy, Equity and Co-education. Buckingham: McGraw-Hill/Open University Press.

Matthews, B., & Sweeney, J. (1997). Collaboration in the science classroom to tackle racism and sexism. *Multi-cultural Teaching, 15*(3), 33–36.

Salovey, P., & Sluyter, D. (Eds.). (1997). *Emotional development and emotional intelligence. Educational implications.* New York: Basic Books.

Sharp, P. (2001). *Nurturing emotional literacy.* London: David Fulton.

Singh, B. (2001). Dialogue across cultural and ethnic differences. *Educational Studies, 27*(3), 343–355.

Steiner, C. (1997). *Achieving emotional literacy.* London: Bloomsbury.

NOTES

[1] This research was funded by the Gulbenkian Foundation and Goldsmiths College.

[2] Full details of these procedures, along with typical tasks given to the pupils, are detailed in Matthews (2003).

DISCUSSION ASSESSMENT **APPENDIX** DA1

Observer: ____________ Class: ________ Date: ________

Name	Each time a person talks, put in a ✓.

Main comments: When you fill in the chart below, do *not* use ticks or numbers. Make a comment like: The most; a lot; the least; frequently; well; not at all.

Name	talking	listening	interrupted others	helped others	how much did they learn?

Other comments
____

INDEX

Science & Technology Education Library

Series editor: William W. Cobern, *Western Michigan University, Kalamazoo, U.S.A.*

20. P.J. Fensham: *Defining an Identity. The Evolution of Science Education as a Field of Research.* 2003 ISBN 1-4020-1467-8
21. D. Geelan: *Weaving Narrative Nets to Capture Classrooms.* Multimethod Qualitative Approaches for Educational Research. 2003 ISBN 1-4020-1776-6; Pb: 1-4020-1468-7
22. A. Zohar: *Higher Order Thinking in Science Classrooms: Students' Learning and Teachers' Professional Development.* 2004 ISBN 1-4020-1852-5; Pb: 1-4020-1853-3
23. C.S. Wallace, B. Hand, V. Prain: *Writing and Learning in the Science Classroom.* 2004 ISBN 1-4020-2017-1
24. I.A. Halloun: *Modeling Theory in Science Education.* 2004 ISBN 1-4020-2139-9
25. L.B. Flick and N.G. Lederman (eds.): *Scientific Inquiry and the Nature of Science.* Implications for Teaching, Learning, and Teacher Education. 2004 ISBN 1-4020-2671-4
26. W.-M. Roth, L. Pozzer-Ardhenghi and J.Y. Han: *Critical Grahicacy.* Understanding Visual Representation Practices in Scholl Science. 2005 ISBN 1-4020-3375-3
27. M.J. de Vries: *Teaching about Technology.* An Introduction to the Philosophy of Technology for Non-philosophers. 2005 ISBN 1-4020-3409-1
28. R. Nola and G. Irzik: *Philosophy, Science, Education and Culture.* 2005 ISBN 1-4020-3769-4
29. S. Alsop (ed.): *Beyond Cartesian Dualism.* Encountering Affect in the Teaching and Learning of Science. 2005 ISBN 1-4020-3807-0